GOOD HEALTH ON A
POLLUTED PLANET

GOOD HEALTH ON A POLLUTED PLANET

A HANDBOOK OF ENVIRONMENTAL HAZARDS AND HOW TO AVOID THEM

Nigel Dudley

Thorsons
An Imprint of HarperCollins*Publishers*

recycled paper

Thorsons
An Imprint of Grafton Books
A Division of HarperCollins *Publishers*
77-85 Fulham Palace Road,
Hammersmith, London W6 8JB

Published by Thorsons 1991

1 3 5 7 9 10 8 6 4 2

© Nigel Dudley 1991

Nigel Dudley asserts the moral right to
be identified as the author of this work

British Library Cataloguing in Publication Data
Dudley, Nigel
Good health on a polluted planet: a handbook of
environmental hazards and how to avoid them
1. Man. Health. Hazards
I. Title
613

ISBN 0 7225 2384 X

Printed in Great Britain by
Mackays of Chatham, Kent

CONTENTS

PART THREE: OUTDOOR HAZARDS

PART FOUR: PESTICIDES

PART FIVE: TRANSPORT

PART SIX: OTHER HAZARDS

ACKNOWLEDGEMENTS

This book has drawn on about fifteen years' experience in the environmental field, working with many different organizations and a large number of individuals. It is impossible to list here everyone who has helped me with ideas, information and comments. Particular thanks are due to my friends and colleagues in London at Earth Resources Research Ltd, Green Audits Ltd and the Pesticides Trust and in Bristol at the Soil Association. Some of the background research contained in sections here was first carried out for the Consumers' Association, Elm Farm Research Centre, Friends of the Earth, Greenpeace and the London Food Commission. People who helped particularly this time with providing information and comments include Malcolm Fergusson, Simon Hodgkinson, Sue Stolton and Jim Sweet. At Thorsons, I am particularly grateful to Kate Allen for initiating the project, to Sarah Sutton for patiently waiting for all my delayed scripts, and to Sara Dunn for sensitive editing. Any remaining errors or omissions are, of course, my own responsibility.

INTRODUCTION

UNDER SIEGE

As I sit down to write this introduction, the French mineral water company Perrier has just withdrawn 160 million bottles of its water after minute traces of benzene were detected in some supplies. This enormously costly gesture is a concrete indicator of the enormous importance now attached to the range of health hazards known as 'environmental contaminants'. Although benzene has been recognized as a powerful carcinogen for years, the trace levels Perrier found would almost certainly have been ignored even up until two or three years ago; at the very most supplies would have been cleared up without destroying the entire existing world stocks. Twenty years ago, beakers of benzene were routinely handed around any chemistry class for students to sniff, and frequently used in experiments.

So what has happened? Perrier's decision is just one result of the enormous loss of public confidence caused by the series of health scandals which rocked different sections of industry throughout the 1980s. It shows just how seriously investors are taking the environmental issue as it relates to our own health. Asbestos, radiation, salmonella and listeria are just a few of the many examples which have paraded almost constantly across the front pages of newspapers in the last few years. Consumer surveys show that people are increasingly concerned about the impact which environmental pollution can have on their health.

All well and good. As an environmentalist, and someone who's spent a good part of the last decade promoting organic food, I'm bound to welcome such a growth in public consciousness. But unfortunately this growth of consciousness isn't matched by a similar growth of understanding about exactly what the hazards, or perceived hazards, actually are. The result tends to be over-reaction to whatever hot potato the media is concentrating on at the time, and continuing ignorance about other issues which may well be more important.

This state of affairs isn't surprising. Unless people are scientifically trained in the particular disciplines needed to understand toxic hazards, they have precious little chance of assessing whether something is really dangerous or not. People who *are* scientifically trained soon find out that the experts usually disagree anyway. So we end up with people carefully avoiding apple juice because of pesticide residues, but still smoking cigarettes – and still eating all the other food containing pesticide residues as well. Of course, the people who prepare the reports for newspapers, radio and television don't necessarily have any specialist knowledge either.

Such ignorance is dangerous. It means that public reaction to important issues is led by sensationalism rather than hard fact. It gives industry the chance to pick holes in criticism and thereby rubbish people who are trying to work towards a healthier environment. It means that people are putting their effort into avoiding some hazards while ignoring others. And it ends up slowing down the whole process of improving things.

This book has been written to try and provide a guide through the maze of health and environment. The introduction gives a background to the issues, a brief guide to how decisions about health hazards are made, how the evidence is collected, who decides what we should be allowed to eat, breathe and drink – and some indication as to whether the whole process works or not!

The main section of the book is an A-Z guide to issues divided into six main sections. Subjects range from additives and asbestos through to visual display units and water. I've looked hard at some of the supposedly 'healthy' options, as well as at known hazards. In each case, a brief overview is given about what the issue is and what we should be worried about, followed by a very broad assessment of how dangerous, or potentially dangerous, the problem is. Lastly there are a few risk-cutting tips and sources of further information. Subjects are cross referenced, and the index gives extra details about locating information.

As we'll see, in many cases there still aren't hard facts available with which to make decisions; people have to judge the evidence as they find it. In almost all cases, there are at least two opposing views. Typically, an industry under attack claims that its particular process is not dangerous, while environmentalists say that it is. There are almost always 'experts' who support both sides in a case. I try to give as fair an assessment as possible, but in all too many cases we still don't know for sure one way or the other.

At the end of the book, there are further resource sections. I have included a glossary, because one of the problems for people trying to make sense of these issues is the mass of scientific terms involved. These are defined when they first appear in the text, and then used and listed in the glossary.

There is also a fairly detailed resource list. This isn't a scientific treatise, and I haven't weighed down the text with detailed references to published papers. But at the same time, I hope it will be of interest to professionals as well as laypeople. Addresses of organizations which may be able to help further in several countries are included.

The basic message of this book is that we should be careful but not paranoid. People live longer today than they did in the past. Some of the 'problems' we will be looking at also have advantages: some food preservatives may cause health risks, but so do many of the moulds which grow on rotting food; if there are questions about the side effects of insulation, there is firm evidence that cold houses kill a good number of elderly people every year.

Beware of buzz words and watch out for people trying to take advantage of the public's fears. All cases need to be judged on their own merits; if not, concern about environmental pollution will be in danger of being treated as a fad rather than a legitimate worry, and people will go back to not caring much again. If that happens, we will all lose out.

WHAT ARE ENVIRONMENTAL HAZARDS?

A couple of years ago I wrote an article on environment and health for a consumer magazine, which contained the seed of the idea for this book. One of the main problems the editor and I had was in deciding exactly whan an 'environmental hazard' was in the first place. Did it include natural hazards, like radon gas and aflatoxin moulds? Or was it confined to hazards created or increased by ourselves, such as pesticides, solvents and Legionnaire's disease? And where did 'deliberate' contaminants like dangerous food additives fit in?

I still can't finally answer that question. But in this book I have assumed that any hazards, or potential hazards, in the home, workplace or just in the open air are legitimate issues for people to be concerned about. You'll find most of the best known issues covered, along with quite a few that you probably have not come across before. It is just as important to know whether walking across a hillside of bracken is dangerous as it is to be able to assess whether you are being slowly poisoned in your office.

Note that this book looks almost entirely at effects on ourselves. With the growth of the green consumer movement, many more people are making consumer choices based on the impact of a product on the environment – for example whether it causes water pollution or helps destroy tropical rainforest. This is entirely laudable, and is a vital part of environmental action. But here we are being more selfish, and looking at the extent to which the effects of pollutants – natural or created by ourselves – can damage our own health.

BUT HOW CLEAR ARE THE LINKS BETWEEN CAUSE AND EFFECT?

To read the headlines in the tabloid press (and sometimes in the more reflective media as well) you might think that any particular issue is an open and shut case: 'Cancer scare food fed to thousands of babies!!!' and so on. In fact, it's very difficult to prove the relationship between any substance and a particular effect on our health. There are two main methods used, and neither of them is perfect by any means.

Toxicological Studies

These involve testing a substance on some other living creature. Traditionally these have been live animals such as rats, mice, rabbits and monkeys. The various campaigns against vivisection argue persuasively that most of the tests carried out are for products which we don't really *need* at all – like toiletries and cosmetics – and that each new product developed causes the death of hundreds or thousands of animals, sometimes in extremely protracted and painful circumstances. Nowadays an increasing number of toxicological studies take place using bacterial cultures instead.

Whether we use a Rhesus monkey or a colony of bacteria in a petri dish, the basic methods are just the same. The test animal is exposed to very high doses of the substance under test, to see whether it develops any disease or other harmful side effects. Once tests have been carried out on hundreds or thousands of animals, a number of broad conclusions are reached. A substance is given an *LD50* value, which is the amount which kills half the population of the test animal, which is an indication of how poisonous it is.

In addition, statistical tests are carried out to show whether a greater proportion of the test animals develop diseases or other side effects as a result of exposure than would be expected in a natural population. The key effects being investigated here are various forms of irritation to eyes, skin or breathing, the development of cancer, detectable genetic mutation, any increase in the proportion of birth defects in descendants of the test animals, and a range of other effects including damage to internal organs, allergic reactions and breakdown of the immune system.

Quite apart from the ethical issues involved, there are serious doubts about the validity of the results of such tests. There are two main flaws. First, all animals react differently, and there is no proof that, for example, something which increases the chance of cancer developing in a rabbit will have the same effect on humans; we know that it's a sign that something is potentially dangerous, and that's about all. When we get to animals nearer ourselves, such as the monkeys or apes, there is apparently an increased chance of a correlation, but evidence that something is an animal *carcinogen* (cancer-causing agent) is certainly not proof that it will be dangerous to us as well.

Second, feeding an animal huge doses of a suspected cancer-causing agent is not the same as exposing it to a low, background, dose throughout its life. The former method is chosen simply because the time and logistic implications of simulating natural exposure make it virtually impossible in most cases. It would take decades to pass anything for use and health specialists have settled for the next best option. This gives people who want to argue that a suspected hazard is not dangerous a lot of room for manoeuvre. You'll often read about industry spokespeople saying that tests are inconclusive and that animal studies are not proof alone. We'll come to a concrete example in the section on nitrates, when laboratory tests on dozens of different animals all suggest that high nitrate levels in the diet cause cancer, while corresponding evidence from studies on human populations suggest that it may not have the same effect. This brings me to the next way of testing products.

Epidemiological Studies

These are very detailed studies of the health of large groups of people, and of all the factors affecting them. For example, if it was suspected that a particular industrial chemical was causing birth defects in the children of people who had been contaminated, an epidemiological study might be carried out on workers in factories using the chemical. The study would have to work out whether birth defects actually were statistically significantly more common among the workforce, and then look at all the other factors which could be affecting them: other pollutants; how many people smoked; dietary patterns; how sedentary the workforce was, and so on. It doesn't take much imagination to see that these studies are hugely complex, difficult to get right, and take a lot of time and money.

There are disadvantages with this method as well – apart from the high costs. First, of course, is that by the time the studies are done, the substance is out in the environment and, if it is dangerous, is already causing harm to people. Second, because it is so complex, there is a lot of room for getting things wrong, and therefore for making the wrong decision or for opponents to argue that the substance isn't really dangerous at all.

Third, the evidence is just *statistical*. This is because we can never *prove* that a particular disease is caused by a particular substance. If someone has worked in a radon mine for years and comes down with lung cancer, as happened to many American Indians and workers in Namibia, we can be pretty sure that their cancers have been caused by the radon. *Pretty sure but never certain*. Statistically, some people get lung cancer naturally, many more people get it from smoking cigarettes, and proving that a particular dust particle of radon started the whole process is not possible.

In practice, most people now accept statistical evidence that something is dangerous if the statistics are strong enough. But the more unscrupulous industries and their representatives still use this as an excuse for doing nothing. So that is why the tobacco industry can say that there is no 'proven' link between smoking and lung cancer or heart disease. The World Health Organization statistics may show that the regular smoker takes five minutes off his or her life with each cigarette, but this evidence is 'statistical'.

SO WHAT ARE 'SUSPECTED' HAZARDS?

Most of the things we will be looking at in this book are suspected hazards. It took decades to get agreement that smoking was bad for health. And let's be clear, smoking is *far* more dangerous than most of the individual issues looked at here. When you read that something is a 'suspected carcinogen', for example, it could mean several things: it could mean that toxicological tests have suggested, or even proved, that it increases the risk of cancer in some laboratory animal species; on the other hand, it may mean that high rates of cancer have been observed in groups of people coming into contact with the substance; or it could mean both.

Governments rely on experts to decide whether a substance is a 'known' or just a 'suspected' carcinogen (or any other type of hazard). The experts often disagree. Governments then decide themselves whether the risk is great enough for it to be banned. Governments often disagree as well, and so substances end up being banned in one country and not in another. Quite often manufacturers continue to make a banned chemical and simply export it to somewhere it hasn't been banned. It often remains illegally available in the banned country as well. In reality, the control of hazardous substances is much more messy than the impression given by smart official handouts.

SO WHY ARE THINGS SO COMPLICATED?

One major reason for the confusion is that there is a fundamental disagreement among specialists about levels of risk. The analysis accepted by most countries is that there is no safe level for substances which cause chronic problems like birth defects, cancer, immune breakdown and so on. Risks may be extremely small with a trace of the toxic chemical, but they still exist. A more traditional approach suggests that there *is* a safe level, below which the substance is totally harmless. This view still prevails among some influential people in the British establishment, and is the cause of much of the confusion in our own handling of pollution issues as they relate to human health.

The notion of safe levels is a dangerous strategy even if you believe it to be true, because how do you make a judgement about what a 'safe' level is? A major misjudgement about safe levels of radiation from nuclear explosions meant that hundreds of British soldiers, and many more aboriginals, were affected by the atomic bomb tests the British carried out in Australia in the 1950s.

Another problem is that although we now carry out careful tests on individual substances, we don't know how they *interact*. Mixtures of some chemicals can increase the effects beyond the sum of their individual effects, i.e. one acts as a catalyst for the other. This is known as *synergism*. There has been virtually no research carried out into the effects of synergism amongst the range of pollutants found in the atmosphere, or water, or food. So we don't have any idea, for example, what it means to be exposed continually to very low doses of a whole range of toxic chemicals all mixed up together – something which we all experience today. Without going over the top and scare-mongering, there are many unresolved issues for scientists to fight about in the hazard field.

OUR OWN ROLE IN ALL THIS

The confusion means that those with vested interests can usually get someone reasonably persuasive to back their views. In cases where health hazards have been suspected, powerful manufacturing groups have an unpleasant history of resisting controlling laws for as long as possible. Virtually all 'hazardous pollutants' have only finally been controlled because consumers, environmental groups, and others have campaigned for a long time – often in the face of public ridicule. (Which is not to suggest that consumers and environmentalists are *always* right!) Perhaps the decision by Perrier, and a few other recent industry initiatives, mean that the days of antagonism are coming to an end, and industry will become more self-regulating. Let's hope so.

SOME THINGS TO WORRY ABOUT

Some of the side effects of pollution will come up so frequently in the book that I will define what they are, and briefly discuss their importance, before we start.

Allergens are substances which stimulate the development of allergies in sensitive people. This can either mean that they cause an allergic reaction themselves, or that they increase sensitivity to other allergenic substances, such as pollen which causes hay fever. Either way, the commonest allergic reaction is something similar to hay fever; a blocked nose, sore eyes and other symptoms of a streaming cold. However, allergies can cause a whole range of other effects as well, including skin problems, stomach aches and sickness. Examples of known allergens include some pesticides and quite a few solvents. Many other materials can cause skin irritation: see, for example, the section on office hazards (see page 127).

Anti-cholinesterase products are chemicals which interfere with the transmission of messages by the nerves. Those affected suffer trembling, weakness and dizziness. People sensitive to anti-cholinesterase substances can sometimes collapse after even mild contamination. Examples of anti-cholinesterase products include many organophosphorus and carbamate pesticides, and some medical products. Sensitivity is well recognized and products usually carry warnings, but the effects of repeated, low doses are less generally recognized.

Carcinogens are things which stimulate the development of cancer. Examples include smoking tobacco; fibres from asbestos; certain kinds of radiation; a number of pesticides; car exhausts, and many more. Carcinogens certainly aren't confined to artificial substances or uses, and powerful carcinogens include wood smoke (probably worse than cigarettes); certain moulds which grow on rotting food; the spores from bracken; and natural sunlight. Proving that something is carcinogenic is particularly difficult. This is also an area where degree of carcinogenicity is very important; although a carcinogen is always hazardous, the level of risk varies enormously between different carcinogens. Unfortunately, we often don't know what the level of risk is in different cases.

CNS effects are effects on the central nervous system, i.e. the brain and spinal nervous system. CNS effects can range from making you slightly drunk or high (like some solvents, which is why people sniff them for kicks) to substances which can cause permanent brain damage. Examples include some forms of insulation, which can cause permanent brain damage if fumes are breathed in during a fire; some pesticides which can cause temporary or permanent CNS effects; and a range of solvents.

Dermatitis is a skin complaint caused by allergens, irritants and other things which make the skin more sensitive. It is also known as eczema. Some people are particularly prone to dermatitis; symptoms are the skin becoming inflamed, dry and flaky. Many solvents, paints and other common household substances can cause this complaint.

Gastro-enteritis is a fairly general term for a stomach upset. Examples of things which can cause stomach upsets include food poisoning of various kinds, swimming in polluted water and so on.

Irritants are similar to allergens, in that they tend to create reactions in the skin, eyes and respiration of people coming into contact with the substance, breathing in any vapour or getting liquid splashed onto their skin. Again, reactions often appear like hay fever or a cold, and are no doubt often passed off as this by sufferers. Examples of irritants include many pesticides and solvents used in DIY and in the workplace.

Mutagens are substances which cause genetic mutations; these can affect us immediately or, more likely, cause problems for future generations. Examples of mutagens which you'll find in this book include some types of radiation; vehicle exhausts, and certain industrial waste products.

Narcotics are capable of making you feel sleepy.

Sensitizers are substances which stimulate allergies in people exposed to them, and are similar to allergens and irritants.

Teratogens are substances which increase the risk of birth defects. Carcinogens, mutagens and teratogens are thought to work in a similar way in the body, but the fact that a substance has one effect does not necessarily mean that it will have others; i.e. something which promotes cancer may also cause birth defects, but not always. Examples of teratogens include thalidomide, a type of drug used to stimulate conception, which has now been banned, and some kinds of pesticides.

Teratogens are dangerous in part because the foetus is very sensitive. The same tends to be true of young children as well, who are frequently less able to withstand environmental pollutants than adults. We will come across several examples of this in the book.

Toxic means immediately poisonous. This is easy to determine through laboratory experiments and, unfortunately, through observations on people who have swallowed the material by mistake. Bottles or packets should be labelled if they are acutely toxic. Something is described as *chronically toxic* if the toxic effects can build up over time. A toxic material which affects babies is known as *embryotoxic* if it can affect an unborn baby in the very early stages of development, or *fetotoxic* if the unborn baby is at a later stage of development.

HOW TO USE THE BOOK

The main part of the book is divided into six sections:

- Food and Drink;
- Home and Office;
- Outdoor Hazards;
- Pesticides;
- Transport;
- Other Hazards.

Each section contains an A-Z of individual issues, each with a brief introduction to the hazard, followed by a presentation of the evidence (or lack of evidence) of the hazardous nature of the subject under review, an assessment, and some ideas for minimizing risk. Each section ends with brief details about further sources of information and contact organizations.

PART ONE:
FOOD AND DRINK

ADDITIVES

The Issue
The possibility of health risks from additives is already well known. Over a hundred of the E numbers used in the European Community have been identified as actually or potentially harmful, although the risks attached to many of these are probably quite small. Countries differ enormously in their legislation on food additives. More worrying is the fact that many flavourings are not tested at all and, despite what you may have been led to believe, they *don't* appear on the label. But are the 'risks' really that significant?

The Facts
Additives are used extremely widely. In Britain, for example, about 200,000 tonnes are added to food every year. Additives are put in food products to fulfil a number of functions. They include:

- Preservatives: used to prevent bacterial attack;
- Anti-oxidants: used to inhibit the spoiling of food exposed to the air;
- Colourings (dyes);
- Flavour enhancers;
- Emulsifiers: to make oil and water mix together (in margarine for example);
- Stabilizers: work with emulsifiers;
- Processing aids: to prevent foods sticking to tins, anti-foaming agents, solvents, polyphosphates (which make food swell with water), neutralizing agents (against acidity) and so on;
- Nutrition additives (mainly vitamin and mineral substitutes).

Some of these are useful. Better preservation of food has undoubtedly reduced the risk of illness – including cancers – by reducing the incidence of dangerous moulds on food (see *aflatoxins* page 29), although nowadays it is possible to store most food safely *without* using additives. But other additives are simply used to boost sales by changing the colour and disguising the quality of generally poor-tasting food, and allowing food processors to increase their profits by stretching a little bit of food out a long way. *Most* additives are added for the convenience of the food industry rather than the safety of the consumer.

We eat a lot of additives. In Britain, it is estimated that everyone eats, on average, 5 kg of additives a year. This is equivalent to about 23 aspirin-sized tablets every day.

In the European Community (EC), many additives have now been regulated, and given a number to allow them to be identified on food labels without writing out the whole name. These are prefixed with the letter E, and are known as *E numbers*. Others, which have yet to be assessed by the EC, are numbered and regulated by individual governments. Still more (the vast majority) are flavourings, which are, in the main, not regulated at all, and are not identified on the label. Similar food labelling occurs in the United States and Australia.

Some of these additives are known, or suspected (that catch-all phrase again!) of having ill-effects on human health. These range from chemicals which are likely to have an effect on people with particular allergies, or a predisposition to certain illnesses, to those which can apparently affect anybody. The fact that an additive is 'natural' – that is, derived from a natural product – is no guarantee that it is safe for humans.

Of course, because additives are valuable to the food industry, proving that something is dangerous enough to warrant banning or restricting is a long and difficult process. Governments vary in the seriousness with which they view the problem. The United States government, for instance, is generally stricter than the British government, and has withdrawn more additives for health reasons. In Britain, the major retail stores have withdrawn several which the government has refused to ban because the retailers want to retain customer confidence.

There is not enough space here to provide a detailed guide to additive hazards – or suspected hazards. However, a few examples will suffice to show the kind of debate about hazards which has now been in process for a number of years.

Amaranth, or E123, is a colouring additive which has been the source of much controversy in some countries. In 1983, it was estimated that 300,000 kg were manufactured in Western Europe alone. Amaranth is added to sweets and soft drinks to produce a red colour. For many years it was incorporated into the blackcurrant drink Ribena, although it was withdrawn in 1985, and the same year the British retailers Safeway announced that they were no longer using it in the own brand-name products. Yet there have been doubts about the safety of amaranth since 1938, and a vigorous debate since Russian studies suggested it was carcinogenic in 1970.

Tartrazine (E102) is the bright yellow colouring used in some fish fingers, cakes, sauces, soft drinks, marzipan etc. It was banned in Norway and Finland after studies linked it to migraines, asthma and hyperactivity in some people. *Red 2G* (128 in Britain) is likely to be found in sausages, sausage rolls and other meat products, soft drinks, cakes, dessert mixes. It also causes hyperactivity and may be mutagenic, and has been banned in over ten countries including the United States, Canada, Switzerland and Japan. Tartrazine and Red 2G are both colourings.

Potassium bromate (number 924 in Britain) is added to bread to bleach it and help produce loaves faster. It was banned in Austria, Luxemburg and other countries after it was linked with stomach disorders and diarrhoea, and it is generally not allowed in foods for children and young babies. *Gallates* (E310–312) are used in dried potato mash, oils, chewing gum and some makes of breakfast cereal. It may damage the liver and irritate the intestine. Some gallates have been banned in the United States.

Additives do not have to be dangerous in themselves to pose a health risk. Flavour enhancers make food taste 'better' than it really is and thus distort eating habits. I put the word 'better' in inverted commas, because what many flavourings actually do is make food taste more powerfully, and thus get people (and especially children) addicted to a range of foods with heavy flavouring-additive concentrations. The use of artificial flavour enhancers, and also of sugar and salt, has encouraged over-indulgence in sweets, crisps and other junk food, and has helped make people in the industrialized countries overweight.

Getting sound advice is difficult. The appearance of articles questioning the safety of individual additives in the press, or of a book of recommendations about specific additives, meets a storm of protest from the manufacturers and food processors affected. There is also a lot of contradictory advice around, and different interpretations, of the levels of risk. For example, in Britain the Soil Association published a pamphlet which collected together information suggesting that over a hundred additives had been identified as actually or potentially harmful. The London Food Commission stated that: '70 per cent of food preservatives, 50 per cent of anti-oxidants, and 50 per cent of approved food colours currently in use are known or suspected hazards'. The British Consumers' Association, more conservatively, published a book on additives which included a list of 58 'which you may wish to avoid'.

This ambivalence extends to countries as well. Something which is regarded as unsuitable in the United States will be used all over Canada, and vice versa. A survey produced for *Earth Report 2*, edited by Edward Goldsmith and Nicholas Hildyard (1990), compared legislation relating to 17 of the more suspect additives in 11 different countries in Europe, North America and Australia. Norway had banned them all, while Britain had banned none at all, with one recommended for withdrawal. Australia had banned four, the United States 11, West Germany six, and so on. (A different choice of additives might well have given a different answer). A lot depends both on the strength of the consumer movement in a country, and on the criteria by which decisions about risks are made.

The United States has probably had legislation in place to regulate *colourings* longer than anywhere else, having introduced controls in the first decade of the century. Norway has regulated colourings since 1935, and has banned artificial colourings in food altogether since 1980.

There are some international bodies which set standards and give advice. These include the International Joint Exposure Committee on Food Additives which is run by the UN Food and Agricultural Organization (FAO), the World Health Organization and the Scientific Committee on Food of the European Community.

Manufacturers have reacted to fears about additives by introducing some 'additive free' lines. To some extent this is a real attempt to avoid those additives with question marks hanging over their health effects, and thus must be welcomed, but some of the labelling is very ambivalent (or deliberately misleading). Some don't remove *all* additives – for example food labelled 'no artificial colourings' may still contain flavourings, preservatives and so on. Common additives, like sugar, are among the most harmful from a dietary point of view, but they do not appear to be so sinister because we all know what they are.

Assessment
There is a lot of disagreement. But medical researchers are increasingly identifying food adulteration, including use of additives, as an important cause of a number of diseases, including cancer.

Minimizing Risk
Remain suspicious! There are increasing numbers of guides to additives on the market for people who want to check up on indi-

vidual chemicals, and some of these are listed below. Most indus-trialized countries now insist on information on additives appearing on the label, so if you take the trouble you can check up. In general, things to watch out for are:

• Highly-processed food, especially if it is brightly coloured;
• Very sweet, or heavily flavoured food;
• Odd reactions after eating particular foods. Watch out especially for signs of hyperactivity in children after feeding sweets or other heavily flavoured foods.

Further Information
The most comprehensive source of information in Britain is *E for Additives* by Maurice Hanssen (Thorsons). This gives a chemical by chemical guide to the additives likely to be encountered. A shorter, pocket version, which just gives an indication of whether there is a possible health risk known or not is *Look Again at the Label* from the Soil Association. In Britain, the Ministry of Agriculture publishes a free booklet explaining the labelling system, called *Look at the Label* (from publications unit, Lion House, Willowburn Trading Estate, Alnwick, Northumberland NE66 2PF).

Other books which provide valuable background information are *Food Additives* by Erik Millstone (Penguin); *Additives: Your Complete Survival Guide* by Geoffrey Cannon and Felicity Lawrence (Century, out of print) and *Food Adulteration* by the London Food Commission (Unwin). For information about children see *Children's Food* by Tim Lobstein (Unwin Hyman).

AFLATOXINS

The Issue
Aflatoxins are toxins produced by some food moulds, which can damage the immune system and are thought to cause liver cancer, along with a range of other health effects. They can grow on a range of foods. There are suspicions that some aspects of modern agriculture increase the risk of aflatoxin production.

The Facts
Aflatoxins are a group of toxic chemicals produced as by-products of some food moulds. It is thought that the main fungus involved is *Aspergillus flavus*. Aflatoxins are believed to be

particularly dangerous, although the fact that they have received attention in the media may have distorted their real importance. They were first identified in the 1960s when about a hundred thousand turkeys died of a wasting disease on a British factory farm, and aflatoxins were identified as the cause.

Aflatoxins are suspected of causing a range of health effects, including both immediate and chronic effects. Although there is still not firm proof, there are strong suspicions, and substantial evidence, of them being carcinogens – causing liver cancer – and mutagens. They are also thought to have an effect on the immune system of some animal species.

Aflatoxins also pose a more immediate health risk, especially but not exclusively in the Third World, where they reach high concentrations in some tropical and sub-tropical countries. One survey found aflatoxins on 80 per cent of foods in an area of Sudan. They can cause a range of poisoning effects and deaths have occurred in India and East Africa, especially among malnourished children. Aflatoxins are also thought capable of causing Reye's Syndrome, hepatitis, Indian childhood cirrhosis, hepatoma and various pulmonary diseases.

A survey carried out by universities in India in 1987 identified aflatoxin poisoning from Taiwan, Uganda, India, Czechoslovakia, New Zealand, Kenya and Thailand. In India, about eighty people died in one incident when contaminated maize was eaten. In Kenya, 12 people died when they ate maize contaminated 12,000 parts per billion with aflatoxin.

Aflatoxins are found on a range of foods, although most of the mould you can see on fruit, jam and so on will *not* be aflatoxins. They are particularly common on peanuts and have been found in some varieties of peanut butter. They are also known to be present in other edible nuts, oilseeds, cereals, pulses, cocoa, spices and dairy products. There are apparently stricter controls on the safety of peanut butter manufacture in the industrialized countries now to avoid this problem.

To some extent, modern food handling has probably reduced the problem, because moulds in general have been reduced. Some authorities identify this as a major, or *the* major, cause in a general decline in stomach cancer incidence in the industrialized North. However, it has also been suggested that heavy fertilizer use, which increases the water content of plants, encourages moulds to develop and may be increasing the problem again. This is still speculation.

Assessment
Aflatoxins are probably not a major problem in Britain, although care should be taken with foods known to be likely to contain them.

Minimizing Risk
There are a number of ways of reducing aflatoxin contamination at source, including manual removal of mouldy looking individual plants, better drying of food, crop rotation to avoid build up year after year, and use of fungicides, although this method can have health problems of its own. In addition, some forms of processing can reduce risk, including heating, exposure to light, hydrolysis with strong alkalis, centrifuging and filtration.

For people living in temperate countries, the following steps can help:

• Avoid any peanuts, maize and other high-risk foods with obvious signs of mould or damp, or which have been stored for a long time outside a sealed container;
• If you buy peanut butter from a small supplier, ask what precautions they take against contamination with aflatoxins;
• It is probably worth being particularly cautious if staying long in a tropical or sub-tropical country, especially in Africa or West Asia. It would be worth getting local advice about the problem in these cases.

Further Information
The most comprehensive reference on aflatoxins I have found is *Aflatoxins in Foods and Feeds* by D. K. Salunkhe, R. N. Adsuk and D. N. Padule, published by the Metropolitan Book Company in 1987 (Metropolitan Book Co Pvt Ltd, 1 Netaji Subash Marg, New Delhi 110002, India). Another detailed text, although now slightly out of date, is *Aflatoxins: Chemical and Biological Aspects* by J. G. Heathcote and J. R. Hibbert (Elsevier).

ALCOHOL

The Issue
Alcohol isn't particularly healthy stuff at the best of times. However, apart from any inherent effects drinking has on our health, there are a number of other problems concerning the way beers and wines are produced.

The Facts

Estimates for safe levels of alcohol consumption have been progressively reduced over the last few years, and many doctors now believe that there is *no* 'safe' level at all. However, most people continue to drink alcohol, although alcohol-free drinks are becoming more widely available, and socially acceptable, in industrialized countries. There are four possible problems associated with alcohol consumption which fall within the scope of this book:

* The *synergistic* role of alcohol with other toxic substances, leading to long-term health problems such as the development of cancer;
* The effects of certain intentional *additives* used in alcohol as preservatives, dyes and stabilizers;
* Contamination of alcohol with agrochemicals such as pesticides at the stage of growing hops and grapes;
* Contamination by other material added at the time of drinking, such as lemons.

Synergism with Other Toxic Materials

Alcohol is not, by itself, thought to be a carcinogen, although slight increases in certain kinds of cancer have been noted in regular drinkers. However, alcohol acts synergistically with tobacco in promoting cancer. This means that if you drink alcohol at the same time as *smoking* a cigarette or pipe you are more likely to get cancer than if you just smoke, even though alcohol is probably not itself a carcinogen. This is bad luck, because the one time when all smokers want to light a cigarette is when they drink, in a bar or at home after a meal. Bars are also places where smoke hangs in the air, so non-smokers can be affected by *passive smoking*. Alcohol may also have synergistic effects on other pollutants, but these links have not been very carefully explored.

Additives

A lot of wine, beer and spirits also contain substantial quantities of deliberate *additives* (see page 25). Since the scandal of some German and Austrian wines being sold containing anti-freeze, consumers have been more aware of potential problems. Additives can include:

* *Preservatives* such as sulphur dioxide and other sulphites (see page 33); tannic acid, added once the wine is made to stop it

going off; and benzoic acid, added to beer;

- *Dyes* including caramel, which is suspected of having health effects, and is used in many dark ales and stouts to give them a richer colour; and potassium ferrocyanide, used in red wine;
- *Stabilizers* such as citric acid, an acidity regulator in wine; potassium tartrate, potassium bicarbonate and calcium carbonate used to deacidify some wines; sorbic acid to slow down the reaction of yeast; and copper sulphate to reduce sulphide levels;
- *Carbon dioxide* to make cheap wine artificially bubbly;
- *Clarifying or fining agents:* including edible gelatin, isinglass, egg white, dried blood powder, bentonite clay, sturgeon's air bladders, silicon dioxide, kaolin, charcoal, carrageen, silica solution, cellulose powder, tannin and many more;
- *Burtonizing agents:* used in British beers to make the water used taste more like that found in the Burton upon Trent area. A dozen or more additives are used, ranging from calcium chloride to sulphuric acid.

In Britain, alcohol differs from food in that manufacturers do not have to declare additives on the label. Most people drinking stouts probably don't realize that the colour is unnatural.

There is particular concern about *sulphur dioxide* and other *sulphites.* Sensitive people can experience nausea, itching, swelling or severe asthmatic attacks and these symptoms are known to have directly caused at least six deaths in the United States. More generally, it is believed to be one of the main agents in causing headaches, nausea, wheeziness and tightness in the chest; i.e. the classic hangover. Sulphites are added to preserve wines once they have been made and stored.

In Britain, the Consumers' Association carried out tests on a range of wines and found the sulphur content near the maximum permitted. Tests showed that people regularly drinking a couple of glasses of most white wines, or three glasses of most red wines, could exceed their 'Acceptable Daily Intake' (ADI), set by United Nations agencies, for sulphur dioxide. (In practice, most people will be taking sulphur dioxide in with other foods as well.)

These figures may be underestimated. Sensitive people are thought to react to amounts of sulphite as small as one milligram. The World Health Organization has recommended that intake of sulphites should be limited to around a third of a milligram per pound of bodyweight a day – which would be about 45 milligrams for a 10-stone person. It has been estimated that a typical glass of red wine can contain 40 milligrams, i.e. almost equal to the WHO limit.

Both the US Food and Drugs Administration (FDA) and the UK Ministry of Agriculture have recently reviewed safety levels for sulphites and are considering cutting the amounts allowed in food. Some US states insist that sulphites are labelled. This is the case in California, so that any Californian wines which do contain sulphites should say so on the label. In Europe, sulphur dioxide can be identified on food by the number E220 and other sulphites by numbers E221 to E227.

Pesticide Residues

In addition, people are becoming increasingly worried about the possibility of high *pesticide residues* (see page 174) in alcohol, especially in wine. Because grapes are often grown in such large vineyards, and have been refined by breeding for so many centuries, they are susceptible to disease. Pesticide applications are often extremely high. Few studies have been made of the residue levels in bottled wine, but some residues are likely to remain.

Pesticide residues are also a possible problem when fruit is added to drinks, such as lemons added to gin or other spirits. Most citrus fruits are likely to have been sprayed both before and after harvest and the alcohol can easily dissolve pesticide residues from the rind. Some pubs and bars now have lemon which has not been sprayed post-harvest.

Assessment

The health effects of the alcohol are probably worse than the health effects of the contaminants! But if you have felt very ill after drinking a small amount of alcohol, such as getting sick or a splitting headache after a couple of glasses of wine, you may well have been affected by a few unintentional residues, or be sensitive to sulphur dioxide.

Minimizing Risk

Apart from the obvious step of not drinking, there are a range of *organic* (see page 59) wines and a few organic beers on the market which avoid many of the problems discussed above. Organic wines are grown with minimal use of a few pesticides, and have, at most, a quarter of the sulphur dioxide added to conventional wine. The range of organic wines is increasing quickly and they are available in shops or by mail order.

Further Information
Thorsons Organic Wine Guide by Gerry Lockspeiser and Jackie Gear, (Thorsons) and *The Organic Wine Guide* by Charlotte Mitchell and Ian Wright (Mainstream) give details about organic wines and also discuss the problems of conventional wines. Details about some of the additives used can be found in *The Residue Report* by Stephanie Lashford (Thorsons). In Britain the Soil Association also has details of organic wines and lists of suppliers.

ARTIFICIAL SWEETENERS

The Issue
Many people use artificial sweeteners as an alternative to sugar for slimming purposes. Diabetics cannot eat much sugar and must either go without sweet foods or find an alternative. There has been a lot of controversy about various artificial sugar substitutes – are they a safe bet for the sweet-toothed weight-watcher?

The Facts
The search for a cheap and non-fattening alternative to sugar has occupied the time and money of the food industry for well over a hundred years. And the controversy about side effects or artificial sweeteners has gone on for almost as long. Anyone who thinks that concern about health effects of foods has only surfaced in the affluent years of the twentieth century may be surprised to learn that saccharin was first banned for health reasons in Germany in 1898! There isn't space in a book of this type to recount the whole complex history of the debates surrounding artificial sweeteners; in this section the key points relating to some of the most common are summarized fairly briefly.

The best known sweetener in the world is *saccharin*, which was invented in the latter part of the nineteenth century, patented in 1885, and almost immediately became the focus of a controversy which continues today. It is 300 times sweeter than sugar. A number of countries, including Spain, Portugal, Hungary and the United States as well as Germany, banned or restricted the use of saccharin in food before the First World War, and it tended to be sold as a pharmaceutical product for diabetics. Ironically, it was sugar shortages in the two World Wars which spurred a relaxation of the bans and, in 1918, the large Monsanto company in

America rapidly expanded sales. In Britain, official attitudes have always been more relaxed, and saccharin has never been controlled.

In the 1970s, the health debate began again in earnest when laboratory tests suggested that saccharin caused bladder cancer. However, the epidemiological studies which followed didn't give such clear evidence, allowing governments which didn't want to ban saccharin, for commercial or political reasons, an excuse to avoid doing anything. In 1977, Canada banned saccharin outright. The United States has not renewed the lapsed ban of 1912 but insists that warning labels be used on all foods containing saccharin. Other countries, including France, Greece and Portugal, have banned the sweetener from food and drinks, although it is still sold in tablet form for people who particularly want it, or for diabetics. The British government defied EC wishes and simply insists that saccharin is identified on a label, but doesn't have a warning. The UK Committee on Toxicology accepted that saccharin is an animal carcinogen, but failed to bar its use because of the equivocal epidemiological studies.

Another, equally controversial, sweetener is *aspartame*, sold as *Nutrasweet*, which is about 200 times sweeter than sugar. This was not invented until 1973, and was introduced into food mainly during the 1980s, after long and bitter legal and scientific battles in the United States. It is used especially in sweets and soft drinks. This has recently been the centre of fresh controversy following articles in a national newspaper.

Here, the debate rests on the quality of the manufacturers' research and safety testing. There are fears that aspartame has a number of effects on the brain, including causing mental retardation and brain lesions. There have been accusations that crucial areas of research were done sloppily, and that inadequate information was used in making decisions about safety in the United States, Britain and, perhaps, other countries which have allowed the use of Nutrasweet. In addition, a letter in the medical magazine *The Lancet* in 1986 linked aspartame with possible epileptic seizures. Aspartame is known to cause problems for the rare people (about 1 in 15,000) who suffer from a genetic deficiency which can cause phenylketonuria, and they are warned to avoid this additive.

There are a number of other sweeteners available, including *acesulfame K* and *thaumatin*, neither of which are allowed for use in ice cream in Britain. Another controversial group are the cyclamates, which were withdrawn in the United States, Britain and in many European countries after health and safety tests suggested unacceptable risks to health. However, the magazine *Private Eye* claimed that this was based on research funded by the sugar industry.

Most of the sweeteners discussed above are intense sweeteners, i.e. they are far more sweet than sugar. There are also a range of 'bulk sweeteners' on the market, which are substances roughly equal to sugar in their ability to make food taste sweet. Some of these, including *sorbitol, mannitol* and *xylitol,* have all been tentatively linked to cancer in laboratory animals, although the British authorities have decided that they are safe to use. When they are consumed in large quantities they can have laxative effects.

Assessment
It does seem that there are still some sweeteners on the market with likely bad effects on health, but estimations of the degree of risk fluctuate wildly. And remember that sugar itself is an unhealthy substance if eaten in excess, and it is, furthermore, promoted by a huge and powerful collection of business interests. Erik Millstone, in *Food Additives* (Penguin) sums up the debate quite neatly when he writes:

> It seems probable that aspartame is a bit safer than saccharin, and both are probably safer than sugar, but that is not necessarily sufficient to encourage us all to consume any of these chemicals.

Minimizing Risks
There are a number of simple steps to take to reduce the risk from artificial sweeteners, some more obvious than others:

- Check the labels. Saccharin and Nutrasweet are identified by law on labels, others may just say 'artificial sweeteners';
- Eat fewer sweet things. Weaning yourself off a high sugar diet is important for many reasons other than just avoiding possible health effects from artificial sweeteners;
- Remember that sweeteners are not just added to 'sweets'. A lot of foods which are promoted as being 'wholesome' and healthy contain an enormous amount of sugar or its substitutes. An average can of baked beans is 15 per cent sugar for example. Yoghurts, muesli bars and other health food or

wholefood shop foods may contain artificial sweeteners;
• It is also worth watching out for 'natural' sweeteners concentrated to unnatural degrees. Some natural fruit juices, for example, are so concentrated that the sugars, which do occur naturally in the fruit, are at much higher levels than you would ever find in nature. Nevertheless, it is getting easier to buy foods without huge amounts of sugar, or other sweeteners, added, although these are generally more expensive.

Further Information
There isn't a single source of information wholly on sweeteners suitable for the general reading public, but several of the books listed in the additives section (see page 25) include details of various specific chemicals. Erik Millstone's book *Food Additives* is useful in giving the political and scientific background as to how some of the most controversial decisions came to be made. There are also a number of good books on the market about how to cook food *without* using large amounts of sugar or any other sweeteners. Try, for example, *Living Without Sugar* by Elbie Lebrecht, (Grafton). Specific information for diabetics is available from health authorities or from associations such as The British Diabetic Association (10 Queen Anne Street, London W1 071-323 1531).

BOVINE SOMATOTROPIN

The Issue
Bovine somatotropin, commonly known as BST, is a *growth hormone* for cattle, created by *genetic engineering* (see page 221). It boosts milk yields by up to 40 per cent. Current plans are to introduce BST in the United States, Britain, other European countries and, eventually, throughout the world. However, there has been an outcry about its introduction and strong resistance from both consumers and farmers, especially in Britain.

The Facts
The introduction of BST has proved almost as controversial as *food irradiation* (see page 45). It could set the scene for the public and political responses to the introduction of many other *genetically engineered* organisms in the future. Farmers' and consumers' groups have looked at both the health and social aspects

of BST, and at the effects it could have on the welfare of cattle.

The large companies involved in the manufacture of BST initially assumed that it would be introduced without much opposition, but this has not proved the case. The public outcry was so large that the European Commission called for a halt on the process of licensing BST to allow more time to study possible side effects. This moratorium will come to an end in 1991 when there is likely to be another tough political battle. And in the United States, the chair of the Senate Agricultural Committee has asked the Food and Drug Administration to carry out an independent assessment of health implications after senior FDA staff expressed worries about its introduction. But, on the other hand, BST is already for sale in some parts of Britain.

Bovine somatotropins are *biosynthetic* versions (i.e. manufactured by biotechnology) of natural bovine growth hormone. If fed regularly to cattle they are said to be able to boost the milk yield by up to 40 per cent although some tests have shown little improvement in yields. To date, most of the arguments about BST have centred on animal rights issues, because it is stressful and uncomfortable for the cow to produce so much milk, and on the social impact of increasing milk yields still further when there is a surplus all over Europe. However, there are unanswered human health questions as well.

In 1988, the London Food Commission published a detailed report on the BST issue: *Bovine Somatotropin: A Product in Search of a Market* by Eric Brunner. The report pointed out that each of the synthetic BSTs is chemically different from the natural hormone, and thus could well have different biological activities. The LFC argued that this suggested:

> the need for separate safety testing and the development of effective residues testing for *each of the synthetic BSTs* in all products derived from BST treated animals before product licenses are granted [original emphasis].

This hasn't happened in practice. Far from it; in Britain milk treated with BST was allowed onto the market during the experimental period, without the public even being told, and without any attempt at labelling.

To date, official and industry bodies have in general given BST the all clear and, in the latter case, promoted it as hard as they can.

For example, the National Office of Animal Health (NOAH) in Britain, the veterinary pharmaceutical's lobby organization, gives unequivocal assurance that 'The use of BST with dairy cows poses no health or safety risks to consumers of dairy products.'

However, scientific disquiet about BST is growing. Dr Richard Lacey of Leeds University has pointed out that the current trials give no indication of possible long-term health effects. He casts doubt on whether BST remaining in milk will invariably be broken down in the stomach, as assumed by the industry, especially in people suffering diseases which reduce the efficiency of digestion. The composition of the milk may be subtly altered. And many hormones speed up the rate and virulence of bacterial infections, which is particularly relevant for cattle during the spread of *BSE* (see page 54).

In an article written for the *Guardian* newspaper in 1989 Professor Lacey stated bluntly:

> BST ... is a potential health risk to humans and its use should not be allowed. Assurances from Ministers that BST ... is safe to drink in milk cannot be justified. The trials, whose whereabouts have not been disclosed and have resulted in some people in the country drinking milk from BST-injected herds, prove nothing about its potential dangers to human health and are now ethically questionable.

In May 1990, new evidence from the US Food and Drug Administration, which casts further doubts on health effects, was leaked to the media. The research was carried out by Monsanto, one of the companies manufacturing BST, but was initially repressed. It suggested that cows injected with BST retain many times the normal amount of the hormone in their milk, and that BST-injected cattle suffered a wide range of abnormalities, including enlarged organs and frequent udder infections.

There are also fears that BST could reduce the efficiency of immune systems in cattle. This would cause stress for the animals, and also lead to higher use of antibiotics, and thus perhaps increased residue levels in milk and meat.

Opposition amongst other organizations is surprisingly strong. In Britain, the National Farmers' Union, Milk Marketing Board, consumers' groups, National Association of Women's Institutes, major dairies, retailers and all political opposition parties oppose the introduction of BST.

Assessment

This is still very difficult, because the results of the tests are not in the public domain. The almost universal opposition from food retailers, farmers and the public, the animal welfare aspects, and the fact that we have too much milk anyway in the industrialized countries, suggests that it would be crazy to introduce it on a large scale.

Minimizing Risk

If BST is introduced, there will certainly be non-BST milk available, although there are currently no plans to label milk as coming from cows which *have* been treated with BST. Any certified *organic* milk should be free of BST and any other artificial growth hormones or unnecessary medication.

Further Information

The London Food Commision report, *Bovine Somatotropin – A Product in Search of a Market*, gives a good overview of the issues, although it is now slightly out of date. It is also worth seeing a paper by Professor Richard Lacey: *Bovine Somatotropin (BT) The Safety Issues*, submitted to The Committee on Microbiological Safety in Food. This was summarized in the Soil Association's journal *Living Earth* issue 170, April–June 1990.

In Britain, the Association of Unpasteurized Milk Producers and Consumers (Path Hill Farm, Goring Heath, Reading, Berks RG8 7RE) can advise about sources of unpasteurized and organic milk. There is also a campaign group The Campaign to Ban Artificial Growth Hormone BGH (Nantygwair, Pencader, Dyfed).

FAST FOODS

The Issue

Fast food chains have come in for criticism from all quarters of the environmental movement, being accused of causing problems ranging from tropical deforestation to malnutrition. There are also a number of specific health risks which are increased by eating some fast foods.

The Facts

Fast food restaurants, cafes and take-aways have been the great catering phenomenon of the 1970s and 1980s, with the huge chains like Wimpy, Burger King and, above all, McDonalds spreading throughout the world. You can now buy fast food hamburgers all over the industrialized world and, increasingly, in the Third World as well. There's a McDonalds two doors down from Mozart's house in Salzburg, and a Wimpy Bar in the centre of Delhi. The fast food chains are now looking towards Eastern Europe as well.

There are detailed and well-rehearsed arguments about the nutritional benefits, or harmfulness, of fast food. A judge in Texas put a stop to some of the more outrageous claims for the nutritional value of the food from some large fast food chains, legislating against advertising campaigns aimed specifically at children. Surveys show that fast foods such as chips, burgers and hot dogs tend to be high in fat, sugar and additives and low in protein, essential minerals and vitamins. Fine for an occasional 'treat' perhaps, but certainly not as a regular part of your diet.

But are fast foods particularly unhealthy because of the way in which they are prepared? Fast foods have a number of special risks attached to them, which you should be aware of before buying:

- They tend to be high in *additives* (see page 25). This is because they have to be stored for long periods, and rely on an instant 'high' of a strong taste. Some fast food outlets put sugar in their french fries or chips, hot dogs and meat pies, and the french fries can have over half your recommended daily salt intake *before* you add extra salt;
- Fast foods are cooked very quickly, often in *microwave ovens* (see page 125). This doesn't necessarily destroy all the bacteria, which can cause problems if the food is already contaminated;
- Some fast foods are a microbiological minefield! Food which is kept warm, and uncovered, for long periods is particularly likely to produce food poisoning. So kebab houses, where reconstituted meat turns slowly on a spit for hours on end, and Chinese or Indian takeaways, if meat or other dishes are kept warm throughout the day, may be high-risk areas;
- Fast food sandwich bars may make up all their meals at the beginning of the day. Sandwiches made with meat, eggs, mayonnaise and so on can, if not properly cooled or protected, build up dangerous levels of bacteria by late afternoon or evening;
- There are also a lot of *reconstituted foods* sold at some of the

fast food stores. These include burgers made out of reconstituted scraps of meat. If these include brains (from any animals) they must be one of the highest risk foods from the point of view of *BSE* (see page 54).

Assessment
On the whole, you are probably better off going to the larger fast food chains from a hygiene point of view. Standards tend to be relatively high, because of the knock-on effect that any scandal about food poisoning would have on other outlets. Hygiene is likely to be less of an issue than the general effect fast foods have on a balanced diet. There are also, increasingly, attempts to improve the nutritional quality of fast foods and to offer choices with fresh vegetables, salads and so on.

Minimizing Risk
If you eat in smaller shops it is probably worth getting to know some in your area which appear to be hygienic, with cleanly dressed staff who don't handle money and food at the same time. Places where you can watch the food being cooked are good.

Further Information
An excellent source of nutritional and political background to British fast foods is to be found in *Fast Food Facts* by Tim Lobstein of the London Food Commission, published by Camden Press.

HEALTH FOODS

The Issue
Health foods are a huge industry today, often confused in people's minds with wholefoods, *organic foods* (see page 59) and so on. Are health foods really the way to good health?

The Facts
Health food shops sell a large range of foods, medicines and vitamin supplements, so it isn't possible to make an overall assessment. However, there are some general points to bear in mind when buying health food.

Health food may not be free of, or even particularly low in, *pesticide residues* (see page 174). A report from the British Ministry of Agriculture, Fisheries and Food Pesticides Residue Working Group, published in 1989, found pesticide residues in virtually all nuts, pulses, grains and honey sold in health food shops. This isn't surprising because many of these crops are grown in the Third World countries where pesticide use is even less well-controlled than it is in the industrialized North. Because we often continue to export pesticides which have been banned in the industrialized countries due to their health effects, there may well be even more toxic residues on pulses and other export crops grown in the South. This re-importing of our hazardous exports is known as the 'circle of poison'.

Facts like these do not necessarily mean that health foods are *higher* in pesticide residues than other food. Avoiding pesticides altogether is impossible in an industrialized country. But for people wanting to minimize their contact with pesticides *organic foods* (see page 59) are a better bet, and are available at many health food stores. Pesticide residues are very likely to be found on *brown bread*, because the bulk of pesticides occur on the husk and shell. Many people are eating brown bread for health reasons, because of the extra roughage, but if the grain is grown conventionally it may well have higher pesticide residue levels. (And even *organic foods* can also contain some pesticide residues, see page 59.)

Perhaps more importantly, there are some health food shops which sell a large range of vitamin supplements and alternative medicines. *Some* of these are not adequately tested and likely to do you more harm than good. Some bogus cures for AIDS have been flooding the market recently, but many less serious illnesses are also the focus for many cures, good and bad. While not wishing to criticize the many honourable practitioners of traditional or alternative medicines, it is important to bear in mind that not all hazardous substances come from large transnational companies!

It is also worth noting that a lot of products in health food shops may be high in sugars. A survey by the British Consumers' Association found that a standard serving of fruit yoghurt contains about five teaspoons of sugar and a bowl of packaged muesli about two and half teaspoons. Some foods sold as specifically 'low sugar' may be sweetened by *artificial sweeteners* (see page 35) and it is always worth checking the label.

Some ostentatiously healthy products may not be very nutritious either. In 1989 the London Food Commission carried out a survey of cereal bars. According to author Tim Lobstein:

> When it comes to fat, many cereal bars are no better than Mars Bars and the amount of minerals and vitamins in them are not much higher than in chocolate digestive biscuits and currant buns.

The Commission analysed over a hundred different muesli bars and found that not one met the UK government Food Advisory Committee's high fibre criteria of 6 grammes for a muesli bar. They also contained high levels of sugar, and one carob bar tested had 47 per cent more sugar than traditional chocolate bars. There are a number of important questions about *mineral water* which are discussed on page 77.

Assessment
Most health food shops try their best to serve up high quality, healthy food. A high fibre diet, with plenty of fresh unrefined food and with the minimum amount of additives has been repeatedly shown to be a positive aid to health.

But beware of charlatans, and don't assume that just because something comes from a health food outlet it is necessarily chemical free or medically effective. This said, with the exception of bread, where it may be worth buying certified organic if you are worried about residues, most of the foods in health food shops are unlikely to be *more* heavily contaminated than in ordinary greengrocers.

Further Information
For information about pesticide residues in health foods, and organic foods, see *Report of the Working Party on Pesticide Residues 1985-1988*, Food Surveillance Paper number 25, from the Ministry of Agriculture, Fisheries and Food in Britain (HMSO).

IRRADIATION

The Issue
Food irradiation uses short bursts of radiation to 'sterilize' food so that it can be stored for longer, and shipped further, while still looking fresh. Despite a huge push to get it introduced by the

food and nuclear lobbies, persistent fears remain about its impact on health and nutrition.

The Facts

Irradiation of food is an issue which is still very much in the melting pot. A few years ago it looked as if irradiation was going to be introduced all over the world, virtually without a murmur, but, since then, opposition has grown steadily. Irradiation is already used widely in some European countries and in the United States, and there is currently a debate about whether EC legislation will be brought in forcing all member states to accept irradiated food.

Irradiation is the use of very large doses of *ionizing radiation*, usually from X-rays or a cobalt or caesium gamma-ray source, to preserve food. Amongst other things, it can:

- Delay the ripening of fruit;
- Kill (or at least render sterile) insect pests infesting food;
- Kill bacteria on food.

At first sight, it offers some impressive advantages over other preservation techniques. Food still looks the same and remains 'fresh' for considerably longer. It doesn't involve adding any chemicals, so can accurately be described as 'additive free'. Irradiation also has a number of other benefits for the food industry, such as improving the baking and cooking characteristics of wheat; increasing the yield of barley and grapes during beer and wine making; and cutting the time needed to reconstitute and cook dried vegetables.

Irradiation has also attracted the backing of many international organizations and national governments. These include the World Health Organization, the Joint Committee of the International Atomic Energy Agency, the Food and Agriculture Organization of the United Nations, the European Community's Scientific Committee for Food and national governments such as those in Britain and the United States.

However, the story doesn't quite end there. Irradiation is bound to have fairly heavy backing because it answers a lot of prayers for the food industry and, even more perhaps, for the beleaguered nuclear industry. If it worked properly, irradiation could solve the bulk of our food bacteria problems, such as *salmonella*, *listeria* and other forms of food poisoning *without* making any major changes to the industry. (Worse still, the food industry might even think it could get away with being sloppier than at present.) Because of the international business pressure, argue the sceptics, some important safety issues have been ignored.

There are four main issues involved in the safety debate:

• Is the food radioactive?
• Are toxic chemicals created by irradiation?
• Does irradiation solve *all* the bacteria problems?
• Is the food still as nutritious?

Is the food radioactive? The first, and easiest, question to answer is about radioactive food. If irradiation is carried out correctly, the food shouldn't remain radioactive. There are radioactive risks involved, but these only really apply to the workers in the irradiation plants (and to all of us when the reactors eventually have to be disposed of).

Are toxic chemicals produced? The question about toxic chemicals is far more difficult to answer. Bombarding food with radiation can alter its chemical structure and there are fears about the creation of toxic chemicals during this process which have still not been satisfactorily answered.

Although a huge amount of research has been carried out, a lot has been proved invalid, either through incorrect use of data or, in one notorious example in the United States, because of deliberate and persistent dishonesty. Scientific data has been used misleadingly and in a biased fashion on many occasions, so that we still don't know for sure one way or the other.

One of the key issues is whether the irradiation process would create chemical mixtures in food which are carcinogenic. Researchers in India claimed to have found a link between potentially cancerous changes in malnourished children who had been fed irradiated wheat. Dr Richard Piccioni, a senior scientist at the New York based Accord Research and Education Association told a US House of Representatives Committee that: 'Scientific literature provides evidence to make a strong presumption of carcinogenicity in some, if not all irradiated food.'

Does irradiation solve the bacteria problem? Despite this being its main selling point, we cannot be at all certain that food irradiation will actually succeed in killing all the bacteria present. It is possible that resistant strains of bacteria will develop. This has already happened in laboratory tests with *Salmonella* bacteria (see page 63).

In addition, barring the development of resistant strains such as those described above, irradiation kills most bacteria present fairly indiscriminately. That is, it kills beneficial bacteria as well as those which cause sickness or disease. Harmful bacteria which remain, or re-infect after irradiation, might thus have a far more open field in which to develop. For example, increased levels of *aflatoxins* (see page 29) have been measured after irradiation in a series of experiments starting in 1973 (although these results are disputed by the United Nations committee which assesses irradiation).

If this is a real risk, it could have very serious consequences. Take the irradiation of chicken. Irradiation would not only kill the salmonella present, but would also destroy the yeasts and moulds which compete with Clostridium botulinum, the cause of botulism. At the doses proposed, botulism would *not* be killed and could continue to multiply. Yet most of the organisms which warn that meat is going bad would be destroyed, leaving the prospect of botulism spreading rapidly in chicken which looks, and smells, perfectly fresh and wholesome.

Is the food still as nutritious? There are also fears about wholesomeness and nutritional value of irradiated food. We are fairly sure, for example, that polyunsaturated fats are damaged by irradiation and it may be that vitamins are reduced as well. Fats in general are changed in taste, and irradiation is not currently being considered for milk products or oily and fatty food.

Many retailers and food experts oppose irradiation. Egon Ronay, probably Britain's best known food writer, wrote in the *Daily Telegraph* in 1989:

> The case for irradiation is definitely 'not proven' and no jury could recommend irradiation beyond reasonable doubt.

Events over the next few years will determine whether food irradiation becomes as widely used as proposed. However, it now looks likely that, if it does, food will have to be labelled as irradiated (which was not in the original EC proposals). This means that the public will be able to choose whether or not to buy irradiated products. Several large retail chains in Britain have already stated that they will not stock irradiated food.

All of the above criticisms have been based on the premise that irradiation will invariably be used within the law. In fact, there are already plenty of examples where it has been used illegally. For example, in 1986 the Imperial Food group illegally irradiated a consignment of sub-standard prawns imported from Malaysia, instead of destroying them as should have been the case. The use of irradiation to disguise poor quality food is a major reason why many food experts have backed the campaign to ban the process.

Assessment
We don't yet know whether irradiation is dangerous or not. The main focus of dissent from consumer groups has been that the decisions being made are fudged through on insufficient or biased data. But we can be fairly sure that, if introduced, irradiation will be used to cover up poor or sub-standard food supplies.

Further Information
The best book available on the food irradiation issue is *Food Irradiation: The Myth and the Reality* by Tony Webb and Tim Lang of the London Food Commission (Thorsons). There is a US version available, *Food Irradiation – Who Wants It?* by Tony Webb and Kitty Smith, and an Australian edition, *Food Irradiation in Australia*, written by Tony Webb and Beverly Sutherland-Smith.

For information in Europe, contact The London Food Commission, 88 Old Street, London EC1V 9AR; in North America contact Food and Water Inc, 225 Lafayette Street, New York MY 10012; and in Asia and the Pacific region contact International Organization of Consumer Unions (IOCU): PO Box 1045, Penang, Malaysia.

LISTERIA

The Issue
Listeriosis is a type of food poisoning, which gained some notoriety in 1989 because the British government blamed soft French cheeses for causing an outbreak. It has gained attention because of its apparent ability to thrive in refrigerators. Is listeria a new 'super disease'? Is it more or less dangerous than *salmonella* (see page 63) and other bacterial infections?

The Facts

Listeriosis is a disease caused by the bacteria *Listeria monocytogenes*. It has been recognized for fifty years, but until recently relatively little has been known about it, or about routes of infection. Compared with *salmonella* and other 'tummy bug' bacterial infections, listeriosis is less common but, when it does occur, much more serious.

Three groups of people are particularly at risk:

- Elderly or infirm people, who suffer a range of effects including blood poisoning and sometimes meningitis. About 20 to 30 per cent of elderly people who get listeriosis die;
- Pregnant women, who themselves experience something like flu, but who may pass the effects on to their unborn children. The babies are liable to be either stillborn or to be severely weakened and die soon after birth. In a study of 34 pregnant women with listeriosis, 16 babies died before or soon after birth and only two were born as well as would normally be expected;
- People who are already ill, and whose resistance is lower as a result.

However, it seems that anyone can become ill if they are exposed to large enough doses of the bacteria. A quick survey of some well-known outbreaks shows that, for the people unlucky enough to get ill, listeria is a lot more serious than milder forms of food poisoning. An outbreak in Massachusetts in 1983 affected 49 people and resulted in 14 deaths. A series of cases in Los Angeles, two years later, affected 142 people and caused 48 deaths, including 30 infants born to women who contracted the disease (and note that, for some reason, pregnant mothers and their unborn children are counted as a single person in these statistics).

Like some other forms of food poisoning, listeriosis is increasing. Official statistics in Britain suggest that incidence has risen from about thirty cases a year in 1970 to about three hundred today. Dr Richard Lacey of Leeds University, who is one of Britain's leading experts on listeria, estimates that real figures may be nearer eight or nine hundred a year, with two hundred deaths. (We have no way of knowing how many pregnant women contract listeria and lose their babies without the disease being identified as anything except flu.)

We are fairly certain that the listeria bacteria are passed to us from animals, but the precise routes of infection have only recently been worked out. A major problem with listeriosis, and the reason why it may be increasing at the moment, is that the bacteria are resistant to a certain amount of heating and also, crucially, to cooling. This means that moderate heat treatment won't necessarily destroy them all, and if a few survive, they can grow again in a refrigerator. This is particularly true if there is an ample supply of easily digestible nutrients.

Although there is still a lot we have to learn about listeria and its transmission, some foods which are particularly at risk have now been identified:

- *Salads* which are pre-prepared and stored in refrigerators along with dressing (which provides valuable nutrients to the bacteria). For example, research in Canada traced an outbreak to coleslaw, which contained cabbage leaves contaminated with listeria from sheep droppings, and where the bacteria multiplied in the refrigerated conditions, fed by the salad dressing;
- *Soft cheeses:* goats' and sheep's cheeses appear to be particularly prone to infection. It also seems that soft cheeses with a hard crust, such as brie or camembert, provide good conditions for bacteria to multiply. 10 per cent of soft cheeses sold in Britain are thought to contain listeria;
- *Salamis and other soft meats* are possible sources of infection. Again, 10 per cent of salami in British shops is thought to be infected;
- *Cook-chill foods* are the most controversial source of listeriosis. These are meats and other foods cooked to destroy bacteria and then rapidly chilled and stored. They are re-heated (rather than re-cooked) later as a convenience food. Slight lack of care during the preparation or storage gives a high risk of infection by listeria, and some experts believe that these are the real cause of the current increase in cases. This was why the British government's claim that French cheeses were to blame caused so much anger. Estimates of the proportion of cook-chill foods which are infected fluctuate wildly from 2 to 50 per cent, although most people nowadays estimate higher rather than lower.

Assessment

Listeriosis is undoubtedly a real problem for people who are at risk, i.e. the elderly, the infirm and, especially, pregnant women.

Minimizing Risk
The simplest way of reducing risks is to avoid the foods most at risk. This means avoiding:

- Cook-chill foods;
- Pre-prepared salads;
- Soft cheeses (especially goats' cheese, sheeps' cheese and soft cow's cheese);
- Salamis and soft meats.

It is also worth taking steps to avoid the breeding of listeria bacteria in your own home, by not creating conditions in which they could multiply. Avoid:

- Storing salad in refrigerators with dressing added;
- Storing soft cheeses and prepared meats for long periods.

Many people won't want to cut out so many good foods. If you are reasonably healthy, and not pregnant, there is probably little danger in eating cheeses and meats. However, cook-chill foods are simply fast foods and it is worth avoiding these altogether.

What Governments Could Do
At the moment, governments are tending to avoid taking listeriosis seriously because they don't want to face up to confronting the powerful interests of the food industry over controlling cook-chill. As long as they persist in dodging the issue, the problems are likely to continue, and may well get worse.

Further Information
Dr Richard Lacey's book, *Safe Shopping, Safe Cooking, Safe Eating* is a good source of information. Details are given on page 127.

MEAT

The Issue
Nowadays a lot of people are getting worried about eating meat for much more selfish reasons than animal welfare or the iniquities of the factory farming system. This section gives a brief overview of where worried carnivores should start reading.

Most of the public attention on meat, at least in Europe, has focused on various forms of adulteration. These fall roughly into two classes. *Deliberate* adulteration includes:

- High levels of *additives* found in many meat products, including *nitrates* deliberately added as a preservative and various forms of *food dyes*;
- The continued use of *animal hormones* outside Europe to boost growth of animals;
- Use of *tranquillizers* to control the behaviour of animals kept in cramped and unnatural housing conditions;
- Plans for the introduction of genetically engineered stimulants such as *BST*.

In addition, there are a number of *accidental* contaminants, in the form of:

- Increased levels of various 'traditional' diseases, such as *salmonella* (see page 63), *listeria* (see page 49) and other kinds of food poisoning;
- The appearance of some new and frightening epidemics, such as *BSE*;
- Other accidental contaminations, obviously linked to our own actions, such as rising levels of some *pesticide residues.*

Meat is treated differently as well, including the introduction of irradiation (see page 45) in some countries, with others soon to follow. And, last but by no means least, the nuclear disaster at Chernobyl, and more chronic nuclear leaks and discharges, means that a proportion of meat may well now be contaminated with *radiation* (see page 231).

Further Information
Some good books are listed under the section on *hormones in meat* (see page 56). At the moment, there are not many groups promoting healthy or humane meat; the debate seems to be between the meat industry and vegetarians and vegans. However, The Soil Association in England (86 Colston Street, Bristol BS1 5BB) has run a 'Safe Meat Campaign' and published the *Safe Meat Guide* to organic and additive-free meat in Britain. The London Food Commission's book: *Food Adulteration* (Unwin) contains much information on different aspects of meat adulteration.

Bovine Spongiform Encephalitis

The Issue

Bovine spongiform encephalitis (BSE) is a disease of cattle which, at the time of writing, is only known to exist in Britain. It is usually called 'mad cow disease' because infected animals suffer from brain disease. It is spreading rapidly and there is a persistent fear that it could be passed to humans.

The Facts

This is a new issue about which there is still a great deal of disagreement amongst experts. Some specialists believe that we are looking at a potentially disastrous plague which could become more important than AIDS; others feel that this is nothing but a local effect on cattle which is of little overall significance.

The first case of BSE was reported in 1985, although it may have been around for some time before that without being recognized. At the time of writing there have been over a thousand reported cases.

The original source of BSE in cows has been traced back to an ancient disease, commonly called scrapie, which affects sheep. This disease was transferred from one species to another when cows were fed concentrates containing brain and other offal from infected sheep. Now, after being confined to sheep for centuries (perhaps even longer), the disease seems to be on the move. A Siamese cat contracted BSE, apparently from its food; zoo antelope have been infected, and mink have passed it between themselves by biting.

So far, the British government maintain that there is 'zero' risk of humans becoming infected from cattle, although senior government scientists are more cautious, saying that the possibility is 'very remote'. The argument is based upon two facts: that humans have never apparently caught scrapie from infected sheep; that we don't eat cows' brains, where the mysterious infective agent apparently congregates.

However, there are several flaws in these arguments:

- The fact that we don't catch scrapie from sheep is no guarantee that we can't catch it from cattle. The infective agent (and we still don't know what it is) may well have mutated, because the disease appears to be more virulent. Cats are immune to scrapie but can apparently contract BSE;

- Although the bulk of the infection is apparently found in the brain, it's very likely that it spreads to other parts of the body as well;
- We *do* eat cows' brains, or have done until very recently. Although BSE was first identified in 1986, it took over three years for the government to ban sales of offal, so infected meat will have been baked in a lot of beefburgers, pasties, sausages and other reconstituted meat before then.

But surely, you may say, animals aren't eaten once they show signs of the disease? This may be true, but unfortunately, it doesn't help much. BSE has an incubation period of about ten years, perhaps more, during which time it is undetectable with current states of knowledge. The only visible signs of disease are a certain sponginess of the brain during the last stages of the disease, and a slight unsteadiness immediately before the animal 'goes mad' and dies. To make matters worse, the government only recently started paying full compensation to farmers for BSE-infected cattle, so before then a good number were sent off for slaughter as soon as they showed signs of a bit of wooziness, so as to get the full market value. Vets and environmental health officers have found frequent cases of this happening.

There is already a human equivalent of this disease, known as Creuzfeld-Jacob Disease. It has a similarly long incubation, is intensely difficult to eradicate and can apparently survive normal sterilization processes on surgical instruments. Cats injected with Creuzfeld-Jacob disease develop symptoms of BSE quite rapidly.

Other countries take the risk much more seriously than the British government and there have been a number of local and national bans, along with considerable in-fighting within the European Community. West Germany and the USSR were amongst the first countries to ban beef imports from Britain, and it is not served in United States airforce bases or by a growing number of education authorities in the United Kingdom. Some experts believe that all cattle from infected herds, currently running at six million animals, should be slaughtered to try and contain the epidemic. To date, the government haven't even laid down regulations for slaughtering the offspring of infected cattle, despite the fact that sheep quickly pass scrapie to their own young.

It may be optimistic to imagine the disease can be confined to Britain. Concentrated food made from British sheep's brains has been sold widely on mainland Europe, so the chances are that the disease has already spread over there, but simply hasn't appeared yet.

There is currently no way of knowing that any particular beef is safe. *Organic cattle* (see page 59), raised to acceptable standards such as those of the Soil Association, are not fed offal and should therefore avoid infection. But because of the long incubation period, herds which are now organic could have been infected before the farm converted.

Assessment
We don't know. It may indeed be true that BSE cannot be passed to humans. However, the balance of evidence at the moment doesn't give any particular reason to suppose that we are unique in being immune. And if not, we could have a potentially enormous epidemic on our hands, starting in about 1995. Let's hope the optimists are right.

Minimizing Risk
We don't even really know enough to avoid the risk, short of going vegan. Avoiding beef products is a start, but we can't be sure that the infection will not spread to milk and cheese. Nor can we be certain that other animals are uninfected; beef offal is still being fed to pigs and poultry.

Further Information
This issue is developing very fast at the moment and is being reported in detail by the daily press in Britain. Check the latest theories and information before making up your mind. Some consumers' groups, including Parents for Safe Food, Britannia House, 1-11 Glenthorpe Road, Hammersmith, London W6 0LF Tel: 081 748 9898, are monitoring developments very carefully. There are, at the time of writing, no accessible books on the subject, although these will doubtless be available before too long.

Hormones in Meat

The Issue
Artificial hormones used in meat have caused a whole range of health scares. They have now been banned in Europe, although widespread illegal use occurs and modern use of *genetic engineering* (see page 221) may help some farmers to sidestep this ban.

Artificial hormones are still used very widely in the United States, Australia, New Zealand and many Third World countries. This has caused trade friction between Europe and the United States.

The Facts

Hormones are natural chemical substances which are secreted by glands in the body. They regulate essential bodily functions such as growth, rate of metabolism, milk production, and sexual functions. The functioning of the endocrine system which controls hormones is complex, and still by no means fully understood.

In recent years, the use of artificial hormones has played an increasingly important part in livestock husbandry, both to boost growth and to control the biology of animals to ensure the livestock unit runs at maximum efficiency.

Artificial hormones can be used, among other things, for:

• Making sure that female animals come on heat at the same time, thus making the breeding programme operate smoothly;
• Inducing miscarriages in feedlot heifers which do not gain enough meat to satisfy commercial requirements when in calf; these are known as *abortifactants*;
• Artificially castrating young pigs without surgery, by blocking the development of testosterone, the male sex hormone;
• Artificially castrating (caponizing) cockerels so that they can be fattened for the table;
• Promoting growth; this is by far the largest use of artificial hormones, and steroids are frequently used. Growth promoters often act by tampering with the sexual system of the animal. Returns on investment in hormones, in terms of extra meat value, can be as high as ten to one, so that there are very strong commercial pressures for farmers to use them.

Most of these uses are unacceptable to many people from a moral viewpoint, and undoubtedly cause animals additional suffering by disrupting their natural life cycles. However, there are also a large range of health effects in humans which have been linked to the use of artificial hormones in livestock.

The best known of these concern abnormal sexual development – always something to grab the attention of tabloids throughout the world. The use of growth hormones based on oestrogen has been implicated in premature thelarche (breast development) and other forms of sexual precocity in children in parts of Puerto Rico, where growth hormones were used in pigs and poultry. Girls as young as three or four developed breasts, pubic hair and started menstruating. Most were found to have increased oestrogen levels. A number of the girls also developed ovarian cysts.

Sexual abnormalities are not confined to children. Use of diethylstilbesterol (DES) to caponize cockerels has been blamed for a number of effects on male farmers, including impotence, infertility, breast development and changes in voice register. Female users have apparently had disordered menstrual cycles. Other artificial hormones have been suspected of causing miscarriages in humans as well as animals.

Several growth hormones, and other artificial hormones, are known or suspected carcinogens. The areas of Puerto Rico where children showed precocious puberty also had elevated cancer levels, and effects have been seen in farmers in other countries as well. There may well be other health effects.

At the end of 1988, the European Community banned the use of several artificial hormones, despite fierce opposition from the British Minister of Agriculture. (Britain had, despite its opposition, actually withdrawn the hormones in 1986 due to the EC ban.) The ban promoted a mini trade war with the United States, who objected to the loss of trade. Britain and Denmark tried to reverse the decision at the European Court at the Hague, without success.

In practice, it probably doesn't make that much difference. There is a flourishing trade in black-market hormones and illegal antibiotics throughout Europe and there have been a spate of cases of illegal usage come to light in Britain, for example in the west of England in 1985. These are likely to be the tip of the iceberg, because there are woefully few analyses carried out to pick up hormone use, and the chances of being caught are fairly low.

Assessment
There is still a great deal of debate about the safety, or otherwise, of hormones used in animals. No one denies that catastrophic events have occurred in the past, but the proponents of hormones claim that the most hazardous have now been banned, and that stricter security measures would have prevented the tragic events in Puerto Rico from occurring in Britain or the United States. Opponents point to the continued fears about many of

those drugs which remain on the market, the appearance of traces in food and the fact that hormones are sometimes cruel and usually unnecessary.

The introduction of *bovine somatotropin* (see page 38) or BST into Britain would allow farmers to dodge the hormone ban. Although BST is 'officially' to be used to boost milk production, it could also be used to promote growth more generally and would be very difficult either to prevent or to detect.

Minimizing Risk
In Britain and Europe many of the most hazardous materials are supposed to have been withdrawn, although as we have seen they often remain available. In the United States, and in many other countries, most meat will have been reared using hormones.

For people wanting to be fairly sure of avoiding hormones, there are sources of additive free meat in many countries, although prices are higher. There is also a growing market in *organic* meat, which does not have any unnecessary hormones or antibiotics, as well as being produced organically.

Further Information
James Erlichman's book, *Gluttons for Punishment* (Penguin) gives a good and readable introduction to the whole issue of hormones in meat, and much more besides, although is now slightly out of date. Sources of hormone-free meat are discussed in *The Born Again Carnivore: The Real Meat Guide* by Sue Mellis and Barbara Davidson (Optima). Sources of organic meat in Britain are listed in *The Safe Meat Guide* from the Soil Association (86 Colston Street, Bristol, BS1 5BB), and *Thorsons Organic Consumer Guide* (Thorsons).

ORGANIC FOOD

The Issue
The market for organic food is booming. But is it worth the extra price? Is it really completely chemical free?

The Facts
Organic food is food produced by organic farming methods. This is, contrary to general belief, not just a return to 'old-fashioned' farming methods. Although it certainly incorporates elements of

traditional farming practices, including rotations, care of the soil, green manuring and so on, it also relies on modern knowledge of ecology, soil science and crop breeding to ensure that pest damage is minimized and soil fertility maintained.

Research shows – again in opposition to the myths put about by the agrochemical lobby – that productivity does not fall dramatically if a farm converts to organic methods and uses them correctly. A report called *Alternative Agriculture*, published in 1989 by the US National Academy of Science, suggests that there need be little loss of productivity under an organic system. Research by Elm Farm Research Centre and the Aberystwyth Institute for Organic Husbandry and Agroecology confirms that this is also the case in Britain.

Organic farming has a number of environmental advantages, in terms of reducing air and water pollution, preserving natural habitats on farms and averting soil erosion. Organic farming systems also incorporate far higher standards of animal welfare than many conventional farming methods. However, from the perspective of this book, the more important questions relate to the quality of the food produced.

Food produced organically is likely to have lower levels – probably far lower levels – of pesticide residues than food produced with pesticides. All pesticides are banned from organic farms – except a few plant-based ones – and the few which are permitted cannot be used on a routine basis. They are also non-persistent and thus break down quickly after use.

However, it cannot be said that all organic food is completely clear of pesticides. Organic farmers can suffer *spray drift* (see page 187) from conventional farms and roadside treatments. Persistent pesticides can remain in the soil in residue form for years, so that fields that have long been converted to organic methods may retain small residues which can appear in food. And organic food which is stored next to conventional food can sometimes be contaminated by pesticides used in storage which 'migrate' from conventional to organic food, usually by evaporating and resettling. This can be a particular problem for grain. Therefore, organic organizations have been careful not to say that organic food is 'chemical free', but that it has substantially less chemicals.

Organic meat and dairy products do not have hormones added to boost growth. This is becoming less of a selling point in the EC, where artificial growth hormones have been banned in meat, but is still important in the United States and other countries. Eggs don't have dyes added to make them look browner, or the yolks more yellow, as is often the case with battery farmed eggs. Animals are not given routine doses of antibiotics, which may be passed to humans in meat. However, if an animal is sick it is treated with antibiotics as necessary to reduce suffering or act as a cure. The philosophy is to cut out all unnecessary use without causing discomfort, pain or ill-health to the stock.

Animals managed under an organic system will not be given feeds based on brain or other material from animals, which reduces the risk of developing the brain disease *BSE* (see page 54) also known as 'mad cow disease'. However, because of the long incubation period, animals which have been converted to an organic system can sometimes have been infected many years previously. Although risks are reduced, they are not eliminated.

In addition, organic food will not have unnecessary *additives* (see page 25), including biotechnology additives such as *BST* (see page 38), nor will it be treated with *irradiation* (see page 45). It will tend to be processed as little as possible and will generally have lower sugar contents.

Many people buy organic because they believe that it tastes better. Blindfold testings have been fairly clear on this point for some products, such as meat, but have had variable results with vegetables, and there is still a need for a more detailed study of this area. It partly depends on how 'conventional' produce is grown; for example if vegetables have been forced rapidly through heavy use of nitrate fertilizer they may have a higher water content and thus less taste.

Other people buy organic because they believe that it contains additional 'goodness' – nutrients, trace elements and so on. This may be true, and there has been some limited evidence to back this up, but there still needs to be a lot more research carried out in this area.

Assessment

People buy organic food for many reasons, and food quality is one of these. Organic food is likely to be much clearer of pesticide and other residues, but probably not totally clear if grown in an industrialized country because of the widespread contamination. It is more expensive, partly because organic farmers receive a premium. However, the bulk of the increase comes from the inefficiency of distribution costs for what is still a small and scattered supply and, in some countries including Britain, the need to import large amounts of organic food, again from all over the place, to meet supply.

Minimizing Risk

The main risk to consumers comes from buying food which is labelled as organic, but hasn't been produced to organic standards. The best way of ensuring against this is to buy recognized quality labels which have regularly inspected farms and detailed standards for organic production.

Unfortunately, there are a number of competing labels on the market, depending on where you are. In the United States, for example, there are many different labels from both independent organizations and state bodies. In Britain, there are a number of different labels, with the Soil Association Symbol now being the best known. Many other countries have more than one certification scheme. There is an EC-wide standard being adopted, which all organic standards within the European Community will have to adhere to in time, which will help get rid of confusion, but this will probably be different from standards used in the United States and Australia.

Further Information

There are numerous books about organic production. For information about getting in touch with organic groups throughout the world, contact the International Federation of Organic Agriculture Movements, c/o Oekzentrum Imsbach D6695, Tholey Theley, Germany, which has 80 member groups and produces an international directory.

In Britain, the Soil Association produce a set of *Regional Guides* (to South East, South West, Midlands, Wales and Scotland) to shops that sell organic produce (from Soil Association, 86 Colston Street, Bristol BS1 5BB) and Alan and Jackie Gear of the Henry Doubleday Research Association have written *Thorsons Organic Consumer Guide* (Thorsons) which includes a list of producers.

In the United States, there are many different organizations. Natural Food Associates, PO Box 210, Atlanta, Texas is a good place to start, as they publish a monthly magazine *Natural Food and Farming*.

In Australia, Soil Association of South Australia Inc, GPO Box 2497, Adelaide, South Australia, has contacts through the country, a regular newsletter and publications.

SALMONELLA

The Issue
Salmonella food poisoning has become a major cause for concern over the past few years, especially in Britain where controversy over levels of salmonella even caused the resignation of Edwina Currie, Minister of Health. But exactly what is salmonella, and how big a risk is it for people eating a normal diet?

The Facts
Salmonella is the best known form of food poisoning. For many people, especially in Britain, the two are virtually synonymous. Like influenza and the common cold, there are many strains of salmonella (about 2,000 are known) and different strains become important at various times. The current rapid growth in salmonella attacks in Britain is due to an increase in the number of cases involving *Salmonella enteridis*.

Salmonella can be found in any food, although at the moment by far the greatest number of cases come from poultry and eggs. Usually 1 to 10 million bacteria are needed to cause an attack, although as little as a few thousand can cause illness in people who are already unwell, or in young children. (However, no one is immune, and there is no proof for the assertion sometimes made that people get some immunity with age.)

Normally, salmonella poisoning results in between 12 and 24 hours of illness, with the worst period being about six hours into the attack. Symptoms are stomach cramps and diarrhoea (which are also symptoms of many other forms of food poisoning). Occasionally the bacteria can get into the bloodstream causing septicaemia – blood poisoning – which is more serious. In a very few cases, the bacteria also migrate to other parts of the body, when they become extremely dangerous. In Britain, about 250,000 people get salmonella poisoning every year, of whom around a hundred die. Almost all the deaths are of people weakened by age, the very young or those with a previous illness, and are caused by bacteria getting into the blood or other parts of the body.

In Britain, there was a rapid increase in the number of salmonella cases reported during the 1980s. Although this may be partly due to better reporting, as the disease has become more widely recognized, it is mainly the result of a rapid increase in actual cases caused by *Salmonella enteridis*. Numbers of cases caused by other strains have remained relatively constant, but *S. enteridis* cases doubled between 1983 and 1988. There has also been a reduction in outbreaks, where many people became ill, and an increase in one-off cases. This suggests that general standards of hygiene are probably improving, but that many more people are being exposed to the bacteria. Sir Donald Acheson, chief medical officer at the Department of Health, admitted in 1989 that the outbreak could be termed an 'epidemic'.

There are two main sources of salmonella bacteria. The largest source is *poultry*, when inadequately cooked. Because of the way in which chickens and other birds are slaughtered and dressed, which involves a lot of handling, infection can very easily be spread between birds at this stage. It has been estimated that between 50 and 60 per cent of raw poultry carcasses are infected with salmonella in Britain and if these are not adequately cooked the infection can be passed across to us when we eat chicken, chicken products, ducks or other poultry. The only reason that we do not have many more cases of salmonella poisoning is because most people do cook poultry fairly carefully.

The other major source of infection is *eggs*, which, despite being less frequently infected, probably cause more actual cases of salmonella poisoning. Here infection can occur either within the bird's body before the egg is laid, or through cracks and pores in the shell after laying, especially if the egg is covered with droppings. If the egg isn't cooked properly, or is used raw in cooking, infection can pass over to humans.

One of the most important factors in the recent upsurge of cases is that many battery chickens are fed remnants of slaughtered chickens as part of their diet. It seems that this practice greatly increases the risk of chickens becoming infected. We don't know if all the bacteria are destroyed during the preparation of this bizarre chicken feed, but there are certainly plenty of opportunities for *reinfection* afterwards. Much of this feed crosses national boundaries, as it is exported and imported into different countries, further increasing the risk of spreading contamination.

'Free range eggs' are not necessarily safe from salmonella either. A lot of the eggs sold as free range actually come from chickens kept in very crowded conditions, with only limited access to a small 'outside area'. Here, rates of infection can be even higher than in batteries because the eggs are trampled more, whereas in a battery cage they role down the gentle slope of the cage bottom and are collected while they are still clean.

Dr Richard Lacey of Leeds University estimates that one egg in 7,000 is poisoned in Britain. This may not seem very many, but we do eat a great many eggs and the rate of the disease has increased six-fold in Britain during the latter half of the 1980s.

Assessment

Unfortunately, the risk of salmonella is still quite high. Many experts now argue that we should avoid all raw or undercooked eggs.

Minimizing Risks

Some of this is bad news! If you want substantially to reduce the risk of salmonella you will probably have to give up eating some favourite foods. The following list gives an idea of the things to think about when cooking and eating.

- All eggs can contain salmonella. Don't believe any signs or notices which say otherwise. In many countries (e.g. in Britain) all flocks have to be tested against salmonella occasionally, and there is still a high incidence;
- Free range eggs aren't necessarily either cleaner or from birds kept in better conditions; many 'free-range' birds actually live in cramped sheds with a small outside enclosure. If you want *genuinely* free range look for eggs accredited by the Free Range Egg Association (FREGG) or sold under an organic scheme such as that run by the Soil Association in Britain, some other recognizable and respected certification scheme, or buy from a farm you know;
- Eggs should be stored at room temperature in the shop;
- Check eggs in the carton for dirt, cracks or blemishes. Eggs shouldn't have been washed (as this suggests they were previously dirty and increases the risk of infection through the shell at a later stage);
- Cook the eggs well. Experiments at Leeds University found that well-cooked omelettes, scrambled eggs and souffles were generally fairly safe as were hard boiled eggs. Lightly scrambled, boiled or fried eggs with soft centres were liable to contain live bacteria if these were originally present. Poached eggs are very difficult to cook enough to eliminate salmonella.

Frying eggs on both sides helps;

- Avoid raw eggs. This may mean avoiding more good food than you realize. Dr Richard Lacey lists 18 food varieties in his book *Safe Shopping, Safe Cooking, Safe Eating*: egg-flip drink; steak tartare; bearnaise sauce; hollandaise sauce; mayonnaise; mornay; and a whole range of sweets including anything containing home-made ice cream, meringues, mousses, egg-custard sauce and so on;
- Susceptible people, including infants, the elderly, people with low resistance to disease, and pregnant women, may want to avoid eggs altogether, except when they are used in baking.

Of course, most people have continued to eat eggs, and many people are prepared to take the risks of an occasional raw egg product. But you should be aware of the risks, especially to the most susceptible people in the community.

What Governments Could Do
In Britain the government has tackled the egg issue back to front. It has demanded testing from all flocks, but with very little differentiation between huge factory units and back-garden producers selling to the local shop. The price and trouble is insignificant to the former but crippling to the latter. And the basic problems, caused by the factory farming methods themselves, have not been tackled.

Further Information
Dr Richard Lacey's book mentioned above gives a good, clear introduction to this and other issues, and is listed in the section on *listeria* (see page 49).

Two of the most useful organizations to contact in Britain are FREGG – The Free Range Egg Association (37 Tanza Road, London NW3 2UA; 071-435 2596) and Chickens Lib (PO Box 2, Holmfirth, Huddersfield HD7 1QT; 0484-683 158).

SMOKING

The Issue
Presumably anyone reading this book will already be well aware that smoking tobacco is extremely damaging to health. For the record, the World Health Organization estimates that a regular

smoker knocks roughly five minutes off his or her lifespan with every cigarette smoked. Smoking is far more dangerous than most of the more exotic environmental pollutants described in this book. And smoking can also increase risks from *other* environmental pollutants.

The Facts

Numerous studies have shown that smokers are more likely to suffer ill effects from a range of other environmental pollutants, ranging from various kinds of *air pollution* (see page 151) in cities; through *pesticide poisoning* (see page 173) to the effects of *radon* gas (see page 138). There are three main reasons for this:

- Smokers are likely to be in poorer health than non-smokers especially with respect to their lungs and heart;
- Inhaling pollutants through a cigarette, pipe or cigar can change them chemically, and sometimes make them more dangerous. This is because the centre of the cigarette acts as a mini-furnace, and can reach extremely high temperatures, which can affect the chemical structure of the pollutants. (Note how often things are listed as being 'hazardous when burnt' in this book.) This can increase risks from, for example, some *pesticides* (see page 173) and *solvents* (see page 236);
- Some of the toxic substances also act *synergistically* with the poisons in tobacco. This means that the *combined* effects of smoking and other pollutants are greater than the sum of their individual effects. This is apparently true for some *air pollutants*.

Assessment

This is probably the single most dangerous and avoidable thing that most readers of this book will encounter in their everyday lives. The official figure of a quarter of smokers dying of smoking-related diseases is almost certainly an understatement if you include the effects of a generally shortening lifespan, outside of the two particular smoking-related killers of lung cancer and heart disease.

Minimizing Risk

If you are a smoker, try to avoid smoking when working with materials that are likely to be toxic and, especially with those which are highly volatile and/or inflammable.

Further Information
In Britain Action on Smoking and Health (ASH) (5-11 Mortimer Street, London W1N 7RH; 071-637 9843) is a good source of information on tobacco hazards.

SMOKING–PASSIVE

The Issue
Everyone knows smoking is dangerous, but how much risk is there in living or working with people who smoke?

The Facts
Every day, on average, every person alive smokes about half a packet of cigarettes. There are now over a billion smokers in the world, puffing their way through five trillion cigarettes a year, and sending 6 million tonnes of tobacco up in flames. This has direct environmental effects: tobacco is cured by wood fuel in many Third World countries, such as Malawi and Kenya, where it is a serious cause of deforestation.

Smoking is almost certainly the single most destructive thing we do to ourselves, followed closely by alcohol consumption. Tobacco is intensely physically addictive and is linked to a whole range of diseases including, of course, lung cancer and heart disease, but also many other complaints including bronchitis, foetal abnormalities, headaches, coughs and so on.

Tobacco smoke also exacerbates many other environmental pollution problems, in two ways. It generally weakens the body, and acts as a catalyst to other toxic materials, including other forms of air pollution. And the act of smoking itself can render pollutants more toxic, because they are inhaled through a mini-furnace which can alter their chemical composition. For example, breathing some pesticide sprays through a cigarette or pipe can make them more dangerous.

In recent years, the debate has shifted from the tobacco companies arguing that cigarettes were safe to a protracted debate about the rights of smokers to enjoy smoking without busybodies telling them whether they ought to or not. Fair enough, but does the smoker put other people at risk as well?

Passive smokers are people who don't smoke themselves but breathe in smoke from other people's cigarettes, pipes and cigars. They inhale mainly 'sidestream smoke', which is the smoke from the burning tip of the cigarette. This contains more toxic materials than mainstream smoke, which is breathed out by the smoker, because it hasn't been filtered through the smoker's lungs. Sidestream smoke makes up 85 per cent of the smoke in a room, i.e. the vast majority of the toxins from smoking go straight out into the environment.

The balance of evidence suggests that passive smoking is bad for you. In America the US Surgeon General concluded that passive smoking causes small, but measurable, changes in the air passages of the lungs of otherwise healthy adults. In Britain the Health Education Authority states: 'People with allergies often find that passive smoking makes their allergies worse.'

Epidemiological studies of people who live or work with smokers are even more damning. A paper in the *British Medical Journal* in 1986 found a 35 per cent increase in lung cancer among non-smokers living with smokers as compared with non-smokers living with non-smokers. The authors concluded that about a quarter of cases of lung cancer in non-smokers were caused by passive smoking, accounting for up to a thousand extra deaths a year.

A study of non-smokers who had worked with smokers for twenty years or more found that the small airways function of the lungs was impaired roughly the same as in someone who smoked between one and ten cigarettes a day.

Researchers at the Universities of Utah and California found that non-smoking women living with smoking partners were three times more likely to have heart attacks than non-smoking women living with non-smokers.

Children are at risk as well. The British Health Education Authority states that: 'The children of smokers are more likely to get bronchitis, pneumonia and other chest infections than the children of non-smokers.' And just about everyone who works or spends times with smokers can get sore eyes, headaches and coughs as a result of the pollution.

The 39th World Health Assembly concluded that:

> Passive smoking or involuntary smoking violates the right to health of non-smokers, who must be protected against this noxious form of environmental pollution.

Assessment
There is little serious doubt that passive smoking is dangerous.

For people who live and work with smokers, it may well be among the more serious problems examined in this book.

Minimizing Risk

Stand up for your rights in non-smoking public transport areas, in your house and by finding restaurants, cafes and pubs and bars (yes, they do exist!) where smoking is banned or restricted to certain areas.

What Governments Could Do

There is now a very strong argument for cutting out smoking in public places where people can't get away from it. This includes non-smoking areas on public transport, or a complete ban on smoking as introduced by London Underground and domestic airlines in the United States.

Most people start smoking because of peer pressure and a huge advertising budget from the major tobacco companies. Tighter controls on billboard advertising, in magazines and newspapers and on adverts on television, radio and cinema, are all urgently needed.

WATER

Tap Water

The Issue

There is increasing concern about contamination of drinking water in industrialized countries. This includes both 'deliberate contamination' through various additives and accidental contamination from industrial and agricultural pollution.

The Facts

This is an enormously complicated issue, with widespread disagreement about occurrence or levels of risk, the logic of current legislation and of ways to tackle the problem. Regulations are regularly breached throughout Europe and North America, and estimations of costs for bringing water supplies into line with the law are often prohibitively high.

There are two, interconnected problems of water pollution. First is *surface water pollution*, which means contamination of streams, rivers, lakes and reservoirs. This is important but, because of a fairly rapid turnover of water, once the political clout is thrown behind pollution controls it is reasonably quick to clean up. More worrying in the long term is pollution to *groundwater sources* or *aquifers*. Aquifers are important water sources in many areas; for example in Britain about 30 per cent of drinking water comes from aquifers. These have traditionally been very clean, but are now being contaminated by a range of pollutants, including *nitrates* (see page 226), *pesticides* (see page 173) and *solvents* (see page 236) in many areas. In underground sources, changeover time of water is much slower, and we still have an incomplete understanding of how long it takes pollutants to leach down to aquifers. This means that pollution is likely to continue in aquifers for a considerable time after it is controlled at source.

Although several specific pollutants are discussed elsewhere in the book, this section gives a brief overview of what the issues are and how they can be tackled. The major pollutants are examined in turn.

Aluminium

Aluminium is deliberately added in some water treatment plants to make the water clearer. It also increases in concentration in areas where water has become more acid because of *acid rain*, as the acid leaches aluminium from soil. (This is one of the main causes of fish death in acidified waters.) In the last couple of years, fears have been increasing about possible links between dietary aluminium and Alzheimer's disease, which is a brain disorder resulting in premature senility.

Aluminium has been found in the brains of many Alzheimer's sufferers, leading to a suspicion that it could help cause the disease. A number of research projects have found that Alzheimer's disease increases in areas with high aluminium levels in the water. A report published in *The Lancet* in January 1989 showed that when the concentration of aluminium exceeds 0.11 milligrams per litre the risk is 1.5 times greater than in areas where aluminium concentration is less than 0.01 milligrams. Although aluminium in drinking water is only likely to represent a tiny proportion of total aluminium in the body, it may have a disproportionately large effect. About 2 million people in Britain drink water with aluminium levels above those recommended by the EC.

Aluminium can probably also cause more immediate health problems, although again this is disputed by some people. In an incident in Camelford, Cornwall, in 1988, high levels of aluminium sulphate accidentally entered the water supply, resulting in a variety of health problems for people who drank it. These included memory loss, nervousness, irritation and burns, which probably resulted from the acid nature of the aluminium sulphate. The Government initially dismissed these effects as mainly psychological but, in July 1990, information from a number of independent reports suggested that some of the people who had drunk the water have suffered brain damage and memory loss. Anthony Wilson, clinical psychologist from St Lawrence's Hospital, Bodmin, found that 75 per cent of the Camelford group had suffered significant memory losses. Tom McMillan, head of clinical neuropsychology at Morleys' Hospital in Wimbledon, also found evidence of effects, and said: 'Overall, these deficits are mild to moderate, and would be likely to interfere with the ability to carry out day-to-day activities.' However, plans for long-time monitoring of intelligence of children in the area are apparently being shelved.

Bacteria

Despite advances in water treatment, problems of bacteria remain and, in some countries, are increasing again. In Britain, disturbing numbers of parasitic pathogens have been found in water, including the amoeba *Giardia duodenalis,* which causes dysentery, and *Cryptosporidium muris.* A number of outbreaks of disease have already occurred, in Leeds, Ayrshire, Oxford and elsewhere. There are still high bacterial levels in water in several other parts of Europe and influxes of tourists during the summertime can increase levels of risk.

Copper

Some copper can dissolve from pipes, especially if they are used for hot water. This is unlikely to present a problem, although copper pipes can be more dangerous if they are soldered with *lead* (see page 116) as this can increase lead levels in drinking water.

Fluoride

The debate about fluoride in drinking water predates virtually all current concerns about water purity, but has still not been satisfactorily resolved. A certain amount of fluoride is essential to maintaining health teeth. Early research suggested that areas which had naturally high levels of fluoride in water were relatively free from tooth decay, and the Water Authorities started a programme of mass fluoridation, by adding large amount of fluoride to drinking water. This helped the aluminium industry, which was able to transform the sodium fluoride left over from aluminium processing into a commercial product.

However, there have been persistent critics of the programme, from a number of different points of view. Some object on grounds of personal freedom and argue against mass medication; others claim that the programe hasn't worked; and a third group believe that fluorides are linked to a range of health effects, including cancer.

Tooth decay has certainly declined in the industrialized countries since the 1950s, but there is no very clear evidence to link this with mass fluoridation (which doesn't take place everywhere). More likely causes are better education about hygiene, improved diet and, perhaps, addition of fluoride into toothpaste. On the other hand, the studies linking fluoride with cancer are also controversial. It is known that fluoride can cause dental fluorosis (diffuse white mottling of the teeth) and more general fluorosis, which is a crippling disease involving bones becoming thicker but more brittle. There are still no clear answers either way about this issue.

Lead

Lead enters drinking water from old pipes and solder, and will be highest in the morning after water has stood in pipes all night. It is a well-known health risk, especially in soft water areas where it is more soluble, and is discussed in more detail in a separate section (see page 116). The EC is putting pressure on Britain at present to reduce lead levels in water in parts of Scotland, where they have long exceeded European safety limits. Some grants are available from the EC for reducing lead levels.

Nitrates

Nitrates arrive in water from fertilizers, manures, and continual ploughing in intensive farming systems. Levels have increased dramatically in the post-war years and now well over a million people in Britain drink water with nitrate levels above the limits set by the European Commission. Similar problems are occurring throughout Europe and in many agricultural areas of North America. See the section on *nitrates* for a detailed discussion about the health risks debate (see page 226).

Pesticides

Pesticides look set to be the major water purity issue of the 1990s in Europe. The EC has set limits for levels of certain pesticides, including *atrazine* and *simazine*, which many countries are finding it impossible to meet. Fortunately these are of relative low toxicity to humans, although they certainly shouldn't be getting into our water supply. What is perhaps more worrying is that, at present, we do not have the techniques, the number of trained personnel, or the political desire, to test for all the pesticides which may be present. This means that other, more hazardous, pesticides may be building up without our being aware of any problems. In Britain, Friends of the Earth identified 298 water sources as being illegally contaminated with pesticides in November 1988.

Polycyclic Aromatic Hydrocarbons (PAHs)

PAHs may leach from pipes lined with coal tar pitch. Many of this group of chemicals are highly carcinogenic. In Britain, they are sometimes occurring at levels up to 20 times those permitted by the EC.

Solvents

Solvent levels in water are also increasing rapidly and, as yet, there are no European controls over levels. The possible health risks from *solvents* are described in more detail in their own section (see page 236).

Assessment

There are undoubtedly real problems with water. But deciding exactly how important these problems are is extremely difficult. Water in the industrialized countries remains far purer (at least microbiologically) than in most of the Third World. And some of the alternatives, as described below, have problems of their own.

Minimizing Risk

Everyone should be able to obtain basic information about purity of their own water supply from their water company. In some cases (e.g. nursing mothers in areas with high nitrates) alternatives may be available. Friends of the Earth in Britain will give advice about how to go about this if you have problems. However, even when you know the problem, it is not always easy to do anything about it.

People drinking tap water, especially in soft water districts or where there is lead piping, should run their cold taps for several minutes in the morning to get rid of any pollutants which have built up in standing water overnight.

For those who remain worried, there are now a number of ways of tackling the problem, including use of *water filters* and buying in *mineral water*. Both of these are, themselves, fairly contentious and are therefore described in their own sections (see below and page 77).

Further Information
There are a number of useful books available on the water issue. *Down the Drain: Water, Pollution and Privatization* by Stuart Gordon (Optima) gives a general introduction to the issue. *Poison on Tap* by Phil and Frances Craig (Penguin) outlines the main problems facing Britain. It was written as a follow-up to a television exposé, in close co-operation with Friends of the Earth and Greenpeace. FoE have themselves put out a detailed technical assessment of pesticides in water, called *Investigation of Pesticide Pollution in Drinking Water in England and Wales*, by Andrew Lees and Karen McVeigh (1988). This is out of print, but may be available at libraries.

Friends of the Earth (UK, USA and many other countries) and Greenpeace (with groups in most industrialized countries and increasingly in the Third World as well) have both put a lot of work into the water issue. National Pure Water Association (Southern Ashe, Gilbert Lane, Whixall, Whitchurch, Shrewsbury) concentrates entirely on the fluoridation issue in Britain.

Other agencies in Britain include: British Effluent and Water Association, 51 Castle Street, High Wycombe, Bucks HP13 6RN (0494-444544); Scottish Rivers Purification Boards Association: City Chambers, Glasgow G2 1DU (041-221 9600); and the Water Authorities Association (1 Queen Anne's Gate, London SW1 9BT: 071-222 8111).

Water Filters

The Issue
Fears about falling water quality have prompted many people to use water filters in their homes. But are they really any good?

The Facts
There are some definite questions about the safety of *tap water* in

many industrialized countries, although on the whole water is still much less dangerous than in many parts of the Third World, where bacterial levels remain high. People have reacted by buying increased amounts of *mineral water* (see page 77) and by using water filters to remove pollutants from tap water.

There are several kinds of filter:

- Jug filters: these incorporate a filter into the top of a jug. They allow some water to be filtered in advance and used when necessary;
- On-tap filters: these are special appliances fitted to taps, and filter water directly. They usually slow the flow of water through the tap, but allow it to be used as soon as it has been poured out;
- On-line filters: these are fitted directly onto the mains water supply. These usually allow you to have a direct outlet avoiding the filter for fast flow, and filtered water for drinking.

In Britain, the Consumers' Association carried out a range of tests on jug filters. By and large, they found that the filters lived up to the manufacturers' claims, but that you would have to choose carefully if you were interested in abstracting specific chemicals; for example if you wanted to remove all aluminium from the water.

Filters were generally good at removing chlorine, copper, lead, some organic chemicals and zinc. Some also reduced aluminium and manganese levels and most removed iron from hard but not from soft water. A major selling point is that they reduce hardness of water, thereby reducing scaling up of kettles etc (and perhaps improving taste). In practice most were found to reduce calcium, but few reduced magnesium, the two main elements involved in creating hardness. Very few reduced nitrates. Not many did much for the taste of the water either, according to an 'expert panel'.

However, there were a number of problems. First, filtered water is liable to experience bacterial growth (because the chlorine has been filtered out) and should be kept in the fridge, and even then not for more than a few days. Water from six different jugs was all found to have high bacteria levels after a week. Cartridges have quite a short life and have to be replaced fairly regularly, or the system doesn't work. This adds to the costs. By 1989 prices, the British Consumers' Association estimated filtering costs, exclusive of the cost of the apparatus in the first place, as 3–7 pence a litre. The water filter and jug should be thoroughly washed once a week. I haven't been able to find comparative tests for the tap and pipe filters.

Assessment
Water filters do seem a good idea if you are worried about water quality, but won't do much good for taste. They will also not extract everything you might be worried about, and need regular maintenance if they are to function properly. They are likely to be especially useful if you are worried about lead levels or something else which most filters can handle. It is probably worth getting a brand from a well-known shop, where you can talk to an assistant, rather than buying mail order. There are apparently some fairly useless filters on the market.

Further Information
See *Which?*, February 1989, from the Consumers' Association, 2 Marylebone Road, London NW1 4DX, Tel: 071 486 5544.

Mineral Water

The Issue
Sales of mineral water have been booming over the last few years. Environmental groups in Britain, and elsewhere, have claimed that this is because the public are disenchanted with the state of *tap water* in many areas, due to pollution levels. But is mineral water really the worth the extra money?

The Facts
Although some people undoubtedly do buy mineral water because they are worried about their health, a survey by the Consumers' Association suggested that most purchasers choose it simply because it's fizzy. Others prefer the taste. And although mineral water is tested before being sold to the public, there are a number of reasons why it may not be universally as healthy as the bottlers and distributors like to claim.

First, because it is not chlorinated, it will probably contain some bacteria. A small concentration of bacteria is usually no problem, but the Consumers' Association found that when bottled water was kept in shops for some time, the number of bacteria could rise to a fairly high level, before declining again. It seems that if you buy very new *or* quite old water there should not be any problem. But unfortunately there is no way of knowing how long any particular water has been sitting on the shelf.

A more serious risk could come from mineral water containing high levels of radioactivity. Tests run at the Atomic Weapons Research Establishment at Aldermaston, England, found that one (un-named) European mineral water contained radioactivity at levels 19 times higher than World Health Organization recommended maximums. They found British mineral water all containing 'very low' radiation levels and, except for Bath spa water which contains some *radon* (see page 138), well within safety guidelines. High radioactivity in spa waters is well known, and was sometimes regarded as one of the reasons they were supposed to be healthier. Even the figure 19 times WHO limits is well within the range of background radiation to which some people are exposed just because of where they live.

More recently, as reported in the introduction to this book, samples of the mineral water Perrier were found to contain small amounts of benzene, a carcinogen.

Assessment
On the whole, mineral waters are probably less contaminated with pollution of various sorts than tap water from some areas. On the other hand, most people don't drink mineral water in a very logical fashion from a health point of view; for example most people only drink it as a cold drink and continue to make tea and coffee with tap water, although many of the pollutants which could be present will not be destroyed simply by boiling. Probably worth continuing with if you like it.

WILD FOOD

The Issue
A few years ago a lot of people interested in countryside and green issues became enthusiastic about eating 'wild food', i.e. native plants collected from the wild. Is this a good practice or not?

The Facts
Although this is outside the main focus of this book, there are a lot of good general environmental reasons for not picking wild plants for food; there simply are not sufficient number of many

potential food species growing wild to make collecting them acceptable. On the other hand, there are some, including blackberries and raspberries, crab apples, hips and haws, which are abundant enough to eat without diminishing their presence in the wild.

It goes without saying that some wild plants are intensely poisonous, and will kill you far more quickly than most of the artificial poisons or potential poisons looked at in this book. Make sure you get the identification right. I used to know someone who became the first person known to medical science to survive (just) eating a water dropwort. This looks rather like a cow parsley but has a tuber. He mistook it for a pignut, which is another member of the cow parsley family with an edible, nut-like tuber of its own. This was a fairly major error, because the pignut is about a centimetre across and the dropwort tuber like a small potato.

Of more relevance to our subject here, some edible fruits may be contaminated with pollutants of our own making. Blackberries growing near a major road are likely to contain higher *lead* levels than it is advisable to eat (see page 116), and may well have other vehicle pollutants as well. On the other hand, those growing in a hedgerow by a farm, or a managed verge, may have been sprayed with pesticide or herbicide (see page 179).

Assessment

There are relatively few wild foods eaten today. It is certainly worth taking care to avoid hedgerow fruits near roads, or where it is obvious that pesticide has been sprayed. If in doubt in the latter case, it is worth asking the landowner. If you can see that a highway verge has been sprayed, by signs of browned vegetation, wild fruit should certainly not be taken. And it is worth remembering that some forestry owners, especially those managing coniferous plantations, spray all non-crop trees and bushes along the sides of tracks.

Further Information

Richard Mabey's book *Food for Free* (Fontana) is a something of modern classic about native food plants in Britain, although I think he does not lay quite enough stress on the need to conserve wild plants.

PART TWO:
HOME AND OFFICE

AEROSOLS

The Issue

Most people think that as long as aerosols don't contain ozone-depleting chemicals then they're OK. In fact, many aerosols release toxic materials which can damage our health.

The Facts

A lot of attention has been given to the role of aerosols in helping create the '*ozone* hole' over the Antarctic, through the use of chlorofluorocarbons (CFCs) as propellants. The campaign run by Friends of the Earth against CFCs in Britain has been very successful, and most aerosol spray cans now say 'CFC-free' or 'ozone-friendly', which gives the impression that they're completely benign. (Most CFCs will be phased out of aerosols throughout the world in a few years time.)

However, ozone-friendly or not, many of the chemicals used in the cans are not very good for us; by squirting out in a fine spray, they are in an ideal state to be breathed in. This can result in problems for people who are particularly sensitive to chemicals released by aerosols, especially if they inhale them in a confined space. A number of deaths have occurred in this way.

As far as I know, there has never been a comprehensive survey of what is found in aerosol spray cans. However, a survey of some of the spray cans sold by major retail stores, in preparation for this book, found glues containing *solvents* (see page 236) which are known to be or suspected of being carcinogenic; oven cleaners containing unnamed chemicals which were identified by the hazards symbol for 'irritant' and *pesticides* (see page 173) which were irritating to the eyes and skin and potential carcinogens and mutagens.

Assessment

Obviously, exposure is usually fairly limited; but aerosols used in a confined space can build up surprisingly high concentrations of chemical in the atmosphere.

Minimizing Risk

- Avoid aerosols. Whether or not they contains CFCs, aerosols are not particularly good for the environment and are an enormously wasteful packaging system;

- If you really find them useful, make sure that the more hazardous kinds (i.e. pesticides, those containing solvents, or those marked with special hazard symbols) are not used in places where there is little air circulation;
- *Never* squirt them into someone's face (so check carefully which way the nozzle is pointing before you start) and don't use them near young children;
- All aerosols are dangerous if they get too hot, or are punctured. They should be kept away from children at all times.

AIR CONDITIONING AND VENTILATION

The Issue
Many modern offices have air conditioning and ventilation systems, with little access to open air. There are a number of distinct problems associated with poorly-designed air conditioning, which are now being taken increasingly seriously by office managers.

The Facts
Office workers have grumbled for years about air conditioning without ever being taken very seriously. Today, it is being realized that air conditioning can create more problems than it solves.

There are three, interconnected, problems:

- Air conditioning can simply recycle stale air, so that office pollutants, bacteria and other pathogens are maintained at a high level. Research in Britain found that absenteeism due to colds and other minor illnesses was 20 per cent above average in offices with air conditioning systems;
- Air conditioning units can themselves harbour bacteria, and provide them with ideal conditions to live and multiply. Air conditioning machines provide warm – and where humidifiers are used, pleasantly damp – living conditions for bacteria. High levels of bacteria, viruses and even parasitic worms have been found in air conditioning systems. The best known result is *Legionnaires' Disease* (see page 120) but there can also be outbreaks of allergic attacks, known as *humidifer fever*, thought to be caused by *Actinomycetes* virus, although possibly having multiple causes;

- In addition, the air conditioning can add to the general feeling of stuffiness and malaise that many office workers experience, due to a lack of fresh air or rapid air circulation. This malaise is now being taken more seriously than in the past and has led to the *sick building syndrome* theory (see page 145). Research in Britain has found that up to 80 per cent of offices with air conditioning systems have faults in their ventilation equipment which could cause health problems.A survey by the London-based research group Building Use Studies found that sick building syndrome was commonest in air conditioned buildings. However, they didn't find conclusive evidence that buildings with humidifiers had higher rates of sickness, and by no means all air conditioned buildings were among the unhealthiest.

Minimizing Risk

Some of the ways of cutting down problems connected with air conditioning units are described in the sections mentioned above. In addition:

- Air conditioning units should be checked regularly to make sure that they don't harbour obvious mould;
- A variety of sterile humidifers are available, which reduce the risks from bacteria in air conditioning units.

If you are concerned about constant outbreaks of particular health problems among staff in a building (or inhabitants of an air conditioned home) it is worth asking an environmental health officer to check bacterial and other pathogen levels in the system.

Further Information

Many of the sources listed in *sick building syndrome* (see page 147) can give advice about minimizing risks or discomfort from air conditioning.

ASBESTOS

The Issue

Almost everyone reading this book will know that asbestos is dangerous. But is it still a danger today? Or have we eliminated asbestos from most environments in the industrialized world? Are some sorts more dangerous than others? There are still many practical questions involved in managing asbestos, and production continues in some parts of the world.

The Facts

Asbestos is a general name for a group of fibrous mineral silicates. There are three common varieties: crocidolite (blue asbestos); chrisotite (white asbestos); and amosite (brown asbestos).

Asbestos fibres occur naturally. They are mined and separated from the rock in which they are embedded, spun into a cloth, and then mixed with resin or cement so that the material can be shaped. Asbestos is both highly resistant to heat and a poor conductor of electricity. This means that it has been used for an enormously wide range of materials, including roofing slates, insulation, fire-resistant doors and fire fighting suits.

Asbestos fibres range in length from microscopic particles to over a foot (30 cm) long. If inhaled, the fibres lodge in the lungs and the tissue of the bronchial tubes. The result, in the long term, can be development of a crippling and progressive lung disease called *asbestosis* which, in its final stages, can make the victims so breathless that they are unable to do simple exercise like climbing stairs. Asbestos also causes a range of other diseases, including lung cancer, cancer of the gastro-intestinal tract and mesothelioma, a malignant cancer of the inner lining of the chest. The health risks from asbestos are better known than for almost all the other substances discussed in this book.

Asbestos has already been responsible for an enormous number of deaths. In the United States, the official estimates are that the potential death toll from exposure to all forms of asbestos is 2 million. The US Environmental Protection Agency (EPA) estimates that 65,000 people are currently suffering from asbestosis alone, and a further 12,000 a year are dying from asbestos-related cancers. Safer conditions probably won't cause the deaths to tail off until the end of the century, because between ten and sixty years usually elapse between exposure and the onset of disease.

The story of the asbestos industry's reaction to suspicions about the health effects of asbestos is one of the most squalid in the whole history of environmental campaigning. The industry fought against controls for years and, in the United States, tried everything possible to avoid paying compensation to victims. The dangers have been recognized since 1918, when elevated deaths in asbestos miners were noted, and they were apparently routinely refused life insurance in the United States. By 1935 asbestos was linked to lung cancer, and by 1960 was known to cause mesothelioma, but no action was taken in the United States until 1971, and in other countries until some time after that. Many countries still allow production of asbestos today.

Death rates amongst people exposed to relatively high doses were enormous. Studies of people spraying asbestos-based insulation found that 20 per cent died of asbestosis, 20 per cent of lung cancer and 10 per cent of mesothelioma; risks for smokers were a hundred times greater than for non-smokers.

In 1982, Manville Corporation, facing a deluge of claims, filed for bankruptcy, which would have frozen all future suits while allowing it to continue trading. In 1987, Manville proposed compensation of $2,500 million over 30 years for victims, but this was opposed by the victims' support groups, who rightly felt it was not enough (there were 100,000 suits outstanding) and by the corporation's shareholders, who thought it was too much.

The EPA has now announced a total ban on asbestos production, because they say that there is 'no level of exposure without risk'. However, in other countries some types of asbestos continue to be manufactured. For example, in Switzerland the Eternit group have agreed to eliminate asbestos from all cement products for sale in Switzerland and Germany, but not from asbestos piping. In Britain both blue and brown asbestos have been officially banned, but only since January 1986. However, white asbestos continues to be manufactured on the argument that it is safer, although there is ample evidence of its links with mesothelioma.

Many Western-based companies have exported the production and use of asbestos to countries without controls, such as South Korea, India and Mexico. Health and safety standards for the workers in these countries are so low that the death toll from asbestos-related diseases is bound to continue into the future.

The ban on new production in the industrialized countries fails to tackle the enormous problem of asbestos already in existence, and incorporated into buildings, machines and other manufactured goods throughout the world. Here, there is a debate about whether it is best to remove all asbestos, or leave it in conditions where it is entirely encased and not an immediate danger to health. The problem is that removal itself can release enormous amounts of asbestos into the air, and cause more harm than good. In cases where asbestos is exposed or crumbling it is vital to have it professionally removed as soon as possible.

Assessment
All kinds of asbestos are highly dangerous, and should be treated with extreme caution. However, air levels found in offices, schools, hospitals and homes containing some asbestos are, of course, far lower than those to which miners and workers were exposed, so there is no reason to panic immediately you discover some asbestos. Indeed, great deliberation is needed to work out how best to tackle the problem.

Minimizing Risk
Unless you have actually worked with asbestos, risks of contracting disease are fairly low. However, you should certainly avoid all materials containing asbestos. The points below explain the steps which should be taken to minimize risks from asbestos:

- If you live in a country where asbestos is still available in various forms, always avoid buying products or materials containing asbestos;
- If, as is probably more likely for most people, you find asbestos in an existing building, think carefully before doing anything about it. Asbestos which is firmly encased in other materials, and which isn't loose, crumbling or giving off a dust, may well be best left in place. It is always worth getting advice from your local authority health and safety division, or from a reputable firm of asbestos specialists before removing asbestos. As this is a potentially lucrative business, there are a lot of dubious firms in existence, so try to get advice from someone.

Asbestos may occur in a number of different places in buildings. Some of the commonest are listed below:

- In fire doors;
- As lagging around water and heating pipes;
- Sprayed on beams for fireproofing and covered with a suspended ceiling;
- Asbestos base cement sprayed onto walls where it acts as a sound absorbent, or on roofs, in sheds, around water tanks and so on;

- As roofing tiles;
- Old textured paint may contain asbestos;
- Insulating board.

One way of determining if you have a problem with asbestos is by having an air test. This will detect whether there are significant quantities of asbestos fibre in the air inside your home or office. However, you can carry out a preliminary check by seeing:

- How accessible the asbestos is to air movements or areas of high activity;
- What physical condition it is in;
- Whether it is crumbling, flaky or easily broken;
- If it is present in air ducts.

Some *brake linings* on cars contain asbestos. Great care should be taken when removing old brake linings (see page 195).

Further Information
In some countries, free advice is available on tackling asbestos.

For example, in the United States the National Cancer Institute maintains an information programme (7910 Woodmont Avenue, Bethesda, MD 20205). The Occupational Health and Safety Administration (national office: US Department of Labour, 200 Constitution Avenue NW, Washington DC 20216) can also give advice. See also the Centre for Occupational Hazards: (5 Beekman Street, New York NY 10038; telephone (212)-227-6220).

In Britain, you are entitled to advice from your district, city or borough council regarding asbestos, including free inspection of your house and analysis of samples, collection of asbestos waste (sometimes free) and advice. Other sources of advice include the Asbestos Removal Contractors Association, 45 Sheen Lane, London SW14 8AB, 071-439-9231; and the Asbestos Information Centre, Sackville House, 40 Piccadilly, London W1V 9PA, telephone 071-439-9231. Several trade unions are also able to offer advice.

Some of the books listed under *office pollution* (see page 129) are useful here, especially *Office Work can be Dangerous to Your Health* (Fawcett Crest). In Britain, the government publishes *Asbestos Materials in Buildings* available from HMSO. A book which is long out of print, but worth looking out for, is *Asbestos*

Killer Dust by Alan Dalton, published by the British Society for Social Responsibility in Science in 1979. Curwell and March's book *Hazardous Building Materials* (E. and F. N. Spon) is also good (see *insulation* page 97).

BATHROOM HAZARDS

The Issue
A lot of attention has been focused – quite rightly – on the use of laboratory animals to test the safety of cosmetics and other non-essential consumer toiletries. But despite a huge toll in laboratory mice, rats and rabbits, a significant number of health risks remain in the average bathroom. These can be from cosmetics and toiletries, paper products, and some cleaning materials. All are described briefly in the section below.

The Facts
As in other 'general' sections of the book, this section collects together data on a whole range of items likely to be found in the average bathroom, and gives brief background information plus advice about minimizing risks. Unfortunately, many products don't give any details of what they contain on the packet or label, or at best give rudimentary data about the main constituents, so using the information below may be difficult.

Bath Foamers
Some bath foamers contain *formaldehyde*, which is discussed in a separate section (see page 218). They are, of course, not an essential bathroom feature!

Cosmetics
There was a big fuss a few years ago because some cosmetics were found to be potential carcinogens. This is still probably the case, although levels of risk are incompletely known. Others are certainly irritants. In general, cheaper imported cosmetics may well have higher irritant levels than those produced in industrialized countries.

Many cosmetics are based on petroleum products, and many people have a certain degree of allergic reaction to these as a result. Petroleum jelly and mineral oil are two obvious examples, but many other cosmetics are petroleum based. If you seem to have an allergic reaction, it is worth changing to vegetable based products, available at many wholefood shops and some large chains.

A few shops, such as the Body Shop in Britain, now record all ingredients on the packet. An increasing number of producers are using 'natural' products. This doesn't necessarily make them safer than artificial chemicals, and it is always worth asking in the store about possible side effects; some retail chains now give staff training in these issues. It is probably true to say that in the developed countries there are more attempts to remove the irritant and toxic ingredients from cosmetics.

Anti-perspirants

Many anti-perspirants are based on *aluminium chlorohydrate* which is an irritant and can cause severe rashes in people with sensitive skin. Cracked or sensitive skin should not be treated. In addition, this chemical can be irritating to the eyes, so aerosols are probably best avoided. A British teenage boy died from what was described as 'excessive but proper use of an aerosol, in a confined space' in 1988, when the propellant gases caused a cardiac arrest.

Other anti-perspirants are based on *zirconium*, which appears to have a fairly low toxicity. There is also some, currently incomplete, data which suggests that using anti-perspirants at all may be dangerous, in that if the anti-perspirants actually stop perspiration they prevent natural toxins from being excreted from the body, and thus building up where they can do damage. It would be worth watching out for more information on this is the future.

Deodorants do not stop perspiration, and are therefore probably better for your health, but contain fragrances and bactericides which can irritate some people's skin.

Eye Makeup

Some brands of eye makeup imported from Asia apparently still contain *lead* (see page 116). This is extremely hazardous and would be illegal in most countries.

Hair Care

Hair sprays contain aromatic compounds of various kinds, including some which are allergens, sensitizers and have animal carcinogens. Many contain polyvinylpyrreolidone (PVP) plastic resin which can cause or exacerbate a reversible lung disease called thesaurosis. There will probably not be any information about what is inside given on the can. You should always try to avoid getting spray into the eyes and keep away from sensitive skin. For this reason sprays should not be used when you are in the bath, because they will be able to come into contact with the whole body.

Hair Colouring

Dyeing (or tinting!) your hair exposes you to a number of dangerous, or potentially dangerous, products. Hair dyes include *hydrogen peroxide*, a powerful bleach, which is an irritant, especially to the eyes but also through the skin. There are a wide range of other detergents, alcohols and other chemicals which it is best to treat with respect. Ingredients used in some makes of hair colouring are suspected mutagens. *Consumer Reports*, an American consumer magazine, advises against use of hair colouring by pregnant women.

Lipsticks

Some lipstick colourings used in Britain are apparently banned in the United States and suspect chemicals, including saccharin (see page 35) and mineral oil, are included in the ingredients. *Titanium oxide* is sometimes used in pink or opaque shades.

Nail Varnish and Nail Varnish Remover

Nail varnishes usually contain a solvent (see page 236), such as *xylene* and *toluene*. Many nail varnish removers contain *acetone*, which is an irritant to the eyes, nose and throat, and care should be taken to keep it off the skin and away from eyes. The bottle should be kept with the top on to avoid fumes straying in the air, and should never be sniffed. If any does get into the eyes it should be immediately washed off with plenty of cold water, and a doctor consulted.

Perm Solutions

Perming the hair is fairly bad for it in terms of hair health, and some of the perm solution is potentially bad for skin as well. Perm often contains *ammonium thioglycollate* which is a poison and a strong allergen and can cause contact dermatitis. This is a fairly well-known problem for sensitive people, and some perm solutions containing it are labelled, so it is always worth checking.

Shampoos

Some shampoos contain *formaldehyde* which is a skin irritant and allergen (see page 218). Anti-dandruff shampoos may also contain *selenium sulphide* which is an irritant and animal carcinogen. It may be absorbed through the skin, although perhaps only to a significant degree through cuts and grazes. The US National Cancer Institute estimates that substantial exposure of the population to selenium sulphide is 'questionable'. Contact with the rest of the body should be minimized; ideally this means not using it in the bath.

Skin Whiteners
There are skin whiteners manufactured in the Third World which still contain *mercury*, an extremely toxic poison. A British firm were fined for manufacturing skin whitener in 1985 – they claimed they were exporting it to Africa, to countries where there are no controls. It is worth enquiring fairly carefully about health effects before using anything to artificially lighten the skin. Most don't work anyway.

Talcum Powder
Talcum powder is sometimes mined from the same area as *asbestos* and can occasionally be contaminated with asbestos fibres.

Tampons
There are a number of health issues connected with tampons, principally the risk of contamination with *dioxins* and the possibility of them causing toxic shock syndrome. These risks are described in a separate section (see page 241).

Thermometers
Older thermometers will still have mercury inside. This is only a danger if the thermometer is broken, whereupon the mercury should be quickly collected up, without getting it onto the skin or breathing in fumes, and safely disposed of by taking it to a special disposal point for toxic materials or, if none is available, sealing it in a container and throwing it out with the rubbish.

Toilet Cleaners
There are a number of different toilet cleaners on the market. In order to kill germs efficiently most are fairly toxic and should be treated with care, although there are some safer products now available.
  Sodium hydrogen sulphate is sometimes used. This irritates the

skin, mouth and mucous membranes. It should be applied only when wearing gloves and care taken not to inhale any dust. There are also a number of acid bleaches on the market. You should never mix two different types of bleach because sodium hydrogen sulphate mixed with an acid can product chlorine gas which is very toxic, and potentially fatal if a large amount is inhaled or it gets trapped in a small area. (This is not very likely in practice.)

Some *acetic acid* toilet cleaners are sold as being milder. This may be true, but a lot depends on the strength of the acetic acid, and they should also be treated with a fair amount of caution. All toilet cleaners are poisonous if drunk and should be kept out of reach of small children.

Toilet Fresheners

Toilet fresheners hung inside the bowl are sometimes made from *paradichlorobenzene*, which is a chlorinated hydrocarbon and possible animal carcinogen. Vapours are easily absorbed through the lung and can be irritating, although it is probably not a major problem at the concentrations likely to be encountered. It has been implicated in liver disorders in chronically exposed people. It is certainly better for health to keep the toilet fresh by leaving a window open!

Toilet Paper

There is currently a debate about toilet paper, tampons and disposable nappies which have been bleached white, because of possible dioxin contamination. White toilet paper has already been banned in Sweden for this reason. These issues are discussed in more detail under *dioxins* (see page 211). There are now a number of unbleached toilet papers on the market, including some made with recycled paper, so it is easy to avoid the problem if you are at all worried. (This is also certainly better for the general environment.)

Toothpaste

Toothpastes contain a range of *additives*. Some contain s*accharine* (see page 35), *titanium oxide* whiteners and other chemicals. It is very difficult to tell what is in many, because they don't list ingredients on the packet. There is some debate about *fluoride*, found in many toothpastes, and this is outlined in the section on *tap water* (see page 70).

Assessment

It's probably worth bearing in mind that a lot of the non-essential toiletry items have some level of health risk attached, albeit quite small. Whether you use them or not will have to be a trade-off between caution and vanity! The more powerful poisons used to get rid of germs are essential to maintaining adequate hygiene standards, but there are safer options becoming available.

Further Information

There is useful background data on some of these issues in *C for Chemicals* by Mike Birkin and Brian Price (Green Print). *The Green Consumer Guide* by John Elkington and Julia Hailes (Gollancz) also gives some background details and addresses of suppliers of some less toxic alternatives. *Home Ecology* by Karen Christensen (Arlington Books) lists many non-toxic alternatives to cosmetics.

The Consumers' Association in Britain reviews many products in its magazine *Which?* and these assessments now increasingly include health and environmental data. There are also many detailed technical guides to chemical hazards, which can be used for checking things you may have individual questions about. Some of the main ones are listed in the reference section at the end of the book (see page 265).

BUILDING MATERIALS

The Issue

A number of building materials have known health risks. In this section we summarize some of those which aren't discussed in their own section of the book.

The Facts

Some building hazards are already well known. Elsewhere we discuss the risks of asbestos (see page 85), insulation (see page 108) and the impact of wood preservatives (see page 245). Below, a few of the less well known issues are outlined.

Material	Hazards
Abrasives	Lead released from old paint when using abrasives can be especially dangerous because it is in the form of a fine dust which can be inhaled, and is likely to

stay around for some time. See *painting and deco-rating* section (page 135).

Adhesives   Often contain *solvents* which can be inflammable and toxic when inhaled. See separate section for health effects of these.

Cement   Serious burns can result from wet cement touching the skin. The *silicon dioxide* in cement can cause silicosis if breathed in large quantities. *Calcium oxide* (*quicklime*) is an irritant of the throat and can also burn eyes. Bronchitis and pneumonia have been reported in cases of chronic exposure. *Calcium dioxide*, which is an additive, can cause eczema and dermatitis in sensitive people.

Damp-   Hydrocarbon *solvent* vapours in damp-proofing can
proofing   be hazardous to health, causing coughing, irrita-
fluid   tion and headaches.

Fillers   There are a wide range of fillers used in building. Most are fairly non-toxic although they can be irritating to the skin and through inhalation as dusts. Expanding filler can contain isocyanates which are toxic and can cause severe lung irritation in sensitive people.

Grouts   Can irritate the skin.

Mould   Often contain fungicides and should be treated
treatments   with caution as a result.

Putty   Old putty may well contain *lead* and should be treated with caution when removed.

Roofing felt   Contains bitumen which consists of hydrocarbons and their derivatives. Like many petroleum derivatives, it is a possible carcinogen and is poisonous, so should be handled with care. Water from bitumen-treated roofs shouldn't be used for drinking or watering vegetables.

Solder   Contains *lead* (see page 116), tin and antimony – all toxic – and should be treated with great care. The vapours are especially dangerous, because the poisons, and especially lead, are cumulative and can therefore build up to dangerous levels through chronic exposure.

Stabilizers   Water-based stabilizers may contain fungicides. *Solvent* based compounds (usually used for exterior work) are more dangerous and should be treated with greater caution.

Tiles: Can release dangerous fumes when burnt if painted
polystyrene   with oil-based paint.
Varnishes   Many contain *solvents* and exterior products may
                also contain *lead* and fungicides. Both these materi-
                als are hazardous so it is important not to inhale or
                ingest any varnish, and rooms where a lot of varnish
                has been applied should be well aired and not be
                used until it has dried and the solvents evaporated.

Further Information
Hazardous Building Materials by Curwell and March (E. and F.
N. Spon) includes some discussion of these, but tends to concentrate on major building materials such as insulation (see page
108) *C is for Chemicals* by Michael Birkin and Brian Price (Green
Print) gives an alphabetical listing of chemical hazards and
includes a useful chapter on building materials.

DOMESTIC PETS

The Issue
Cats and dogs almost invariably pick up fleas and lice sometimes.
The pesticides used to control these are amongst the most hazardous. In addition, pets can themselves pass on diseases and parasites to their owners.

The Facts
Pet shops sell a range of shampoos, aerosol sprays and flea collars
to prevent or get rid of infestations by fleas and lice. Many of
these contain hazardous pesticides, which are carefully controlled
for use in greenhouses or gardens, but are used fairly indiscriminately on the pets which share our houses, play with children, and
so on. Admittedly, concentrations are likely to be fairly small, but
this advantage is offset by their close proximity to people (and to
pets!). Many people don't wash their hands to eat after stroking
'their' animal on the assumption that it is clean; they will then be
passing any pesticide residues directly into their mouths.

Some of the most hazardous pesticides are used on pets. For
example: *lindane* and *carbaryl* are used in some pet shampoos;
dichlorvos and *diazinon* are found in some flea collars; and
dichlorvos and *fenitrothion* are added to aerosol flea sprays. The
table on page 98 summarizes some of the main health questions

surrounding pesticides commonly used on pets. (Note that not all of these will be available in every country covered by this book, because control over pets products varies a lot.)

Pesticide	Effects
Carbaryl	Very toxic when swallowed, moderately toxic when passed through the skin. Suspected carcinogen and mutagen. Anti-cholinesterase product. Banned in Germany.
Diazinon	Fatal if swallowed, can be absorbed through the skin. Suspected mutagen and teratogen. Anti-cholinesterase compound. Eye and skin irritant.
Dichlorvos	Very toxic if swallowed, absorbed through the skin or inhaled. Mutagen and suspected carcinogen. Anti-cholinesterase compound. Products are labelled in the United States to warn not to use near infants, old people or the sick, or where food is being prepared. The World Health Organization classifies it as 'highly hazardous'.
Fenitrothion	Toxic if swallowed and can rapidly be absorbed through the skin. Anti-cholinesterase agent. Synergistic with dichlorvos. Heavily restricted in the USSR.
Lindane	Toxic if swallowed or absorbed through the skin; wide range of health effects including suspected carcinogen and teratogen. See *wood preservatives* (page 247). *Lindane* is a particular hazard to cats, because they don't seem able to excrete it from the body, so that high levels can build up over time.

In addition, pets can pass on disease to their owners. This is already well accepted in many parts of the world, especially where rabies is endemic, but is often ignored in places where rabies is uncommon or, as in Britain at present, absent. Yet pets can easily pick up a range of diseases from other animals (for example if a cat catches rats and mice), by coming into contact with other pets, sniffing droppings and so on.

Some of these can be serious. For example, faeces from domestic dogs and cats can carry toxocariasis, a roundworm which can affect humans. Research in 1980 found that on average 12 per cent of British dogs were infected, but in some areas prevalence was much more common, rising to 51 per cent in Glasgow. According to the Hospital for Tropical Diseases, up to a hundred children suffer eye damage every year as a result of infection by *Toxocara canis*. The parasite can come into the house on shoes and bicycle and pram wheels, and can survive for several days on the floor. In Britain, it has been estimated that about a hundred tonnes of faeces are left on our streets every day. Pets can carry a range of other serious and not so serious diseases, and can infect their owners with fleas and other parasites.

Assessment

The levels of pesticides used to control parasites in pets are fairly low, but there are some very toxic pesticides on the market. I haven't been able to find any proper assessments of risk to either humans or pets from domestic pesticides. However, it is probably worth trying to avoid use as much as possible, and also to choose the brand quite carefully and take some precautions once the treatment has been given. There are certainly very real risks from infection from pets, and considerable care is needed in some areas to avoid this.

Minimizing Risk

There are a number of steps which can be taken to avoid, or at least reduce, use of pesticides on pets. These include:

* Avoiding build up of infestation in the first place, by:
  —Regular washing and airing of bedding, if possible by putting it through the hot wash in a washing machine;
  —Vacuuming areas where the pet often lies. Dust should then be sealed in an airtight container and disposed of as soon as possible to stop the parasites from escaping again;
  —Regular brushing of the pet.
* Choosing the least toxic pesticide for treatment. For example:
  —Avoiding lindane-based products for cats;
  —Bedding treated with *methoprene* which is probably among the least toxic of the options;
  —There are also some herbal remedies. These shouldn't ever be assumed to be safe, but are probably less toxic than the most potent pesticide formulations. *Pennyroyal* is a traditional herbal fleabane and can be used in fabric flea collars;
  —Apparently *garlic* crushed into pets' food is a useful way of stopping worm infestation.

Note that risks are not just to humans! Almost all the pesticides used can be absorbed across the skin and can damage the health of the pets they are supposed to protect.

* In addition, it is important to ensure that everyone washes their hands thoroughly after handling a pet, and especially before eating. This is particularly important in the case of young children, both because they will be less resistant to disease, and because they tend to suck on fingers and hands, thus increasing risks of contamination.

It is also very important that care is taken to ensure domestic pets do not foul in places where people are likely to be sitting, or gardening. Unfortunately, it is also important to check that contamination doesn't take place in parks and other open spaces. Pets should be trained to defecate in a box or tray, and droppings flushed down the toilet.

Careful worming is needed to eliminate *Toxocara*, and your vet can advise on this. Faeces should apparently be burnt for two days after worming. If your dog (or neighbouring dogs) are known to have had the worm, take great care because it can survive for at least two years in garden soil.

Further Information
There are some good books listing environmentally healthy ways of looking after your pets. For example, the League for the Introduction of Canine Control (PO Box 326, London NW5 3LE) has some useful leaflets. *Home Ecology* by Karen Christiansen (Arlington Books) has a useful chapter on pets.

What Governments Could Do
Countries vary enormously in their attitudes to pets and fouling. Some European countries, for example, have strict rules about controlling animal faeces, and owners are expected to carry a 'pooper scooper' to take away all droppings from paths or grassland. In Britain, regulations are extremely lax, and there is little incentive for pet owners to train or control their animals. Stricter controls on animal fouling should be a priority for health protection.

DRY CLEANING

The Issue
The solvent used in dry cleaning is hazardous. This poses risks for

people working in, or living above, dry cleaning premises. It can also leave residues on clothes.

The Facts

The solvent used in dry cleaning fluid is called *perchloroethylene*, also known as *tetrachloroethylene*. It is moderately toxic if swallowed, absorbed through the skin or inhaled. The vapour is narcotic at high concentrations and can also act as an irritant; repeated exposure can lead to dermatitis. It is also a suspected carcinogen and mutagen. There is a case of enough being passed on through the breast milk of a regular visitor to a dry-cleaning establishment that her child became seriously ill.

Assessment

This seems to be a potentially serious problem for people living above dry-cleaning establishments, and also for people who work in them. Enough solvent can remain on clothes to make it worthwhile taking care when handling them.

Minimizing Risk

- Air dry-cleaned clothes before storing or wearing them;
- Do not carry them in unventilated cars.

DUSTS

The Issue

Dusts are everywhere. Far from just being something you mop up with a cloth, both natural and 'artificial' dusts can have bad effects on our health.

The Facts

'Dust' is a loose term embracing a lot of different substances. The *Concise Oxford Dictionary* defines dust as 'finely powdered earth or other matter lying on the ground or on surfaces or carried about by wind'. In practice we can think of dust in the air in rather the same way as plankton in the sea; dust is made up of a whole range of particles of varying sizes, some living material but mostly inert. What we take for 'clear air' is actually more like a thin soup made up of many different materials. Try looking at a shaft on sunlight and see the visible dust floating; there will be many more invisible constituents as well. These could include:

- Pollen;
- Smoke particles;
- Tiny pieces of earth and rock;
- Small fragments of metal;
- Soots;
- Microscopic organisms including bacteria (dust mites);
- A whole range of smaller particles of chemicals, pesticides, solvents and so on, which are probably not usually thought of as dust at all.

In the United States, the slightly more precise term *fine particles* is used, often shortened to PM_{10}.

The effects of some dusts are well known. Miners, who work long hours in dusty conditions, suffer a higher than average incidence of a whole range of lung diseases, including emphysema, lung cancer, bronchitis and specialized diseases such as silicosis. Uranium miners are perhaps the best known example of this, because of the appallingly high incidence of lung cancer as a result of breathing *radon* dust (see page 138). *Asbestos* (see page 85) miners also suffer particularly badly, but coal and slate miners are also among those who face a considerably shorter life expectancy because of breathing dust.

For the majority of people lucky enough to work in other places, dust problems are generally – but not invariably – less severe. They have been most carefully studied in the office environment, but are also important in factories, for people working by busy roads and so on. Health effects can include the following:

- Breathing problems or pain;
- Lung damage of various kinds;
- Allergies (especially hay fever);
- Some fine particles can also contribute to risks of cancer.

Assessment
It is difficult to separate the impact of dusts from other *air pollutants* (see page 151) in urban areas. Dusts can certainly add substantially to allergic reactions and the general stuffy or unwell feeling typified by *sick building syndrome* (see page 145) but their longer-term dangers are difficult to assess.

Minimizing Risk

For people working in offices, or just living in houses with a dust problem, there are a range of technical and management controls which can limit exposure. These include:

- Keeping surfaces clean through *collecting* dust (with a vacuum cleaner or damp cloth) rather than dispersing it again with a duster;
- Making sure that carpets are cleaned regularly to prevent a reservoir of dust building up in the wool (brushing a carpet disperses dust without reducing it);
- Use of dust control equipment, such as *electrostatic precipitators* and *ionizers*. These must be large enough to service the whole room, and be well maintained, including having the filters regularly changed;
- Installing and maintaining an efficient and adequate *air conditioning system* (see page 84) (even if this is only making sure that a window remains open!).

Further Information

Office Work Can be Dangerous To Your Health by Jeanne Stellman and Mary Sue Henifin (Fawcett Crest) contains information about reducing dusts in the office. See the section on *sick building syndrome* for more general control methods (see page 145).

FLUORESCENT TUBES

The Issue

Fluorescent tubes have had a bad reputation for giving people headaches and even, occasionally, epileptic fits. There have been a spate of scares that they can cause cancer. They also contain a number of toxic chemicals.

The Facts

Fluorescent lights have become popular because they give out about four times the amount of light as a conventional (incandescent) bulb for a given amount of energy. As such, they have clear environmental advantages. Health debates concern both the possible effects of fluorescent light in everyday use, and the risks from toxic materials contained in the tube and surround.

Over the last twenty years, fluorescents have been accused of a number of health effects, concerned with both the emission of ultra-violet light and the 'flickering' nature of the light which is produced.

Ultra-violet light is associated with skin cancer (see *ozone* page 157) and one study published in the British medical journal *The Lancet* made a tentative link between fluorescent lighting and the onset of melanoma, which is the more serious, and often fatal, form of skin cancer. However, the results were certainly not definitive, and most authorities believe that the dose is very low, as most of the UV light is absorbed by the plastic surround. The UV dose would be insignificant compared to that obtained from *sunbathing* (see page 166), for example.

However, the links with more general health effects are stronger. Many people complain that the 'cool white' light given off by the majority of fluorescent tubes contributes to making them feel physically below par in work environments (see *sick building syndrome* page 145). The flicker, which is often made worse by faulty setting, can increase this 'annoyance factor'. It is apparently a problem for some people prone to epileptic attack, in the same way as a flickering television set can be dangerous. Many people complain that it causes headaches and eyestrain. However, there has been no large-scale study to try to quantify these issues in working or living conditions.

There are also potential problems with toxic chemicals. Fluorescent tubes contain small amounts of *mercury*, which is extremely toxic. Great care must be taken to avoid contamination if the tube is broken or cracked. Older models also contained *PCBs* in the capacitors used as starters. These are intensely poisonous, cumulative chemicals.

Assessment
Despite years of research and dozens of exposé magazine articles, there still isn't any very clear information about health risks from fluorescent tubes. Any risks from UV are likely to be tiny compared to lying in the sun or just being out in the open air. However, the role of fluorescents as a factor in sick building syndrome is probably more real, and SBS is now being taken much more seriously than has been the case in the past.

Minimizing Risk
There are a number of simple steps which can reduce any potential problems:

- Ensure that the light is covered with a plastic tube, which will absorb most of the UV light;
- If the very white light is annoying, replace the cover with one which gives out a warmer, yellowish, light (such as those often used above bathroom mirrors);
- Be careful not to break the tube or starter. If you have facilities for disposal of hazardous waste in your area, send old fluorescent tubes there. This service is commonly available in the United States, for example, but very rare in Britain.

Further Information
For a very detailed analysis on the issues surrounding fluorescent tubes, see *Fluorescent Lighting – A Health Hazard Overhead* (1987) by the London Hazards Centre (308 Grays Inn Road, London WC1X 8DS).

FLY SPRAYS

The Issue
Flies are a nuisance and also carry bacteria and disease. However, many of the fly sprays sold contain some of the most toxic pesticides and should be used with great care, if at all.

The Facts
Aerosol fly killers are becoming increasingly popular for use in kitchens, caravans, and other areas where people live. While it is important to control flies, especially in areas where food is being prepared, there are also hazards in the sprays used. In addition, some of the fly killer resins contain toxic materials. Among the pesticides likely to found (which should be listed on the can in most countries) are:

- *Dichlorvos* (in both fly sprays and fly killer resin) – an organophosphorus insecticide and a suspected mutagen and teratogen, irritating to eyes and skin; highly poisonous; behavioural effects have been noted in low doses on rats;
- *Fenitrothion* (fly sprays) – organophosphorus insecticide. A poison and irritant, and a suspected mutagen. It is important to avoid contact with the eyes;
- *Lindane* (fly and wasp sprays) – described in several other places in this book (see *wood preservatives* page 247) – a

poisonous organochlorine pesticide, a suspected carcinogen and teratogen with a range of known and suspected health effects on both humans and animals. Extremely persistent and banned in a large number of countries;

- *Malathion* (professional household pest control) – quite poisonous, irritating and a potential teratogen. There is evidence of emotional effects on people working with malathion;
- *Pyrethroids* (fly spray) – a very variable groups of synthetic compounds, chemically similar to the naturally-occurring pyrethrum. Some are certainly safer than other pesticides on the market, others have suspected health effects of their own;
- *Pirimiphos methyl* (fly spray) – organophosphorus insecticide which is poisonous, irritating and identified as a potential mutagen.

Further health effects of most of these are described in *aerosols* (see page 83), *domestic pets* (see page 97) and *lice and nits* (see page 123). They are all potentially harmful, especially if used near food or in an enclosed space. There are also a range of newer alternatives on the market, including many synthetic pyrethroids. Most of these will be safer than the organochlorine and organophosphate pesticides used more commonly in the past, but all are likely to have potential health effects.

Minimizing Risk

- The first step is making your kitchen (or anywhere else) as unattractive to flies as possible. That means leaving no food lying around uncovered, no rotting bits and pieces in the compost bucket, clean surfaces and tight fitting cupboards. Strong smelling food is especialy attractive to the average house fly;
- Fly papers (i.e. sticky strips of paper which you hang up to attract and trap flies) are fairly effective. But they are also very cruel, as the flies die slowly stuck to the paper. If you do use them, they should be removed fairly regularly and burnt for hygiene's sake;
- There are a whole range of ultraviolet fly traps which are meant to attract flies and then zap them with an electric current. When the British Consumers' Association tried out an assortment they didn't find any of them very effective;
- There are also a lot of folk remedies for getting rid of flies and wasps, including homemade traps which act a bit like fly paper, attracting the flying insects through a little trap-door made of

folded paper inside a jar where they remain trapped. Again, this is probably healthier for us but needlessly cruel to the insects. Some people find it effective to put a small amount of food outside, and some way from the door or window, to attract flying insects away from the house. This sometimes works on picnics as well, so that wasps will tend to stick to a dollop of honey placed away from people;

- If you do need to use a fly spray, there are a few simple rules which can cut down risks:
  —Never use fly spray directly where food is being eaten or prepared;
  —Keep away from infants and young children;
  —Vacate the room for a while after using the spray;
  —Make sure that a sprayed area is adequately ventilated;
  —Pick the safer chemicals; this probably means *pyrethroids* (although some of these are more hazardous) and *pirimiphos methyl*.

FURNITURE

The Issue
Furniture can contain a range of hazardous, or potentially hazardous materials. Some of these are certainly dangerous enough to be worth bearing in mind when making a purchase.

The Facts
There are a number of potential hazards to look out for.

Chipboard
A lot of chipboard, used widely in furniture, doors and interior fittings, contains *formaldehyde*, which can be toxic to susceptible people and may be a carcinogen (although this is hotly disputed).

Polyurethane Foam
This is used as stuffing inside a lot of furniture. It emits cyanide fumes when burnt, and has been a major cause of deaths in domestic fires over the last few years. People can be killed by the fumes entering, for example, a bedroom before they are even aware that their house is on fire.
Since March 1989, all polyurethane foam sold in Britain has had to contain fire-retardants by law. Polyurethane foam contain-

ing fire retardants should be labelled as 'combustion modified high resilient foam', or CMHR.

Plastic
Some PVC plastics emit dangerous fumes, including *phosgene gas* and *hydrogen chloride* in a fire.

Assessment
Toxic fumes from furniture are now a major cause of death in home fires. Day to day problems from furniture are unlikely to be a serious problem, but any possible risks can be further reduced by a few simple strategies.

Minimizing Risks

- Any chipboard likely to contain formaldehyde can be sealed by painting exposed surfaces, thus stopping release of the gas;
- When buying new or second-hand furniture containing polyurethane foam, check that fire retardants have been included;
- In the event of a fire, get out of the house *as quickly as possible*, without waiting to collect valuables, even if the heat is not yet very intense. If there are other people present, seal off the fire by closing doors to slow the spread of heat and fumes.

Further Information
C for Chemicals by Michael Birkin and Brian Price (Green Print) gives extra information on all these chemicals.

INSULATION

The Issue
Over the last few years, various forms of insulation material have come under suspicion of causing for a range of possible health effects.

The Facts
Insulation is important. It is important from an overall environmental perspective, where conservation of energy is going to be an increasingly vital way of combating pollution problems from coal and nuclear power stations, including acid rain, low

level *radiation* (see page 231) and the greenhouse effect. But insulation also has more immediate benefits of cutting fuel bills and allowing people to stay warm. Many old and infirm people die every winter from hypothermia, which could be avoided by domestic energy conservation.

That said, some of the forms of insulation on the market have known or suspected health risks attached, to balance against the advantages. This is one of the more contentious areas covered by the book; the home insulation industry is extremely quick to attack any suggestion that products are at all dangerous, and countries have responded in different ways to the perceived risk. In this section, there is a brief overview of the debate about some of the best known insulation products. Ways of minimizing risk are given under the separate sections.

Insulation is used in a number of different situations: in roof spaces; around hot water tanks; under the floor; within cavity walls; and as cladding on the inside or outside of walls. Because the same materials are often used for more than one purpose, they are listed alphabetically below.

Asbestos
Asbestos was used very widely as a form of insulation in the past, as tiles, a sprayed coating, around water pipes and so on. It is extremely dangerous and is described in detail in its own section (see page 85).

Cellulose Fibres
These are used in roof spaces. They are environmentally sound in that they can be made by re-using waste paper, but they almost always have insecticides and fungicides added for long-term survival, which can cause a health problem. Fibre toxicity hasn't been ascertained as yet, and will depend on the size of the fibres involved, with smaller fibres likely to be more irritating.

Fibreglass
Fibreglass is made up of small pieces of artificial fibre and, like asbestos, is fire resistant and also resistant to many chemicals. It is the type of insulation often used in loft spaces to lay between beams. Fibreglass has been the subject of a lot of debate because it is superficially very similar to asbestos, and there have been fears that it could be as dangerous. There are, indeed, problems with using fibreglass but the levels of risk seem to be much less.

Fibreglass causes health problems because:

- It is extremely irritating to the skin. People laying fibreglass should wear a complete covering of old clothes, including long sleeves and gloves;
- It is also irritating to the upper respiratory tract. Working with fibreglass means you will inevitably breathe in fibres unless you are well protected. You should always wear a mask with a fresh filter when handling or laying fibreglass, and throw away the filter afterwards;
- It can also be irritating to eyes, especially if dust is created. People installing fibreglass should wear eye goggles. Many researchers have found that the irritant effects of fibreglass, and other artificial mineral fibres, are most acute when someone uses them for the first time, or after a long break, so that the occasional domestic user is likely to be at proportionately greater risk of irritation;
- There is some evidence from animal experiments that fibreglass can cause pleural mesothelioma, the very rare cancer of the lining of the lungs which can be caused by asbestos. This, of course, is what most of the health concerns centre on, although the evidence remains ambivalent and there is no similar evidence from studies of human population (and also see below under 'mineral wool'). It seems worthwhile treating it with respect and taking precautions as outlined when using fibreglass, and also avoiding the fibres drifting down from the loft space into the main part of the building once in position.

Mineral Wool

A more general term for a range of fibrous insulation material which includes fibreglass and rockwool. Similar safety precautions are worth taking with all kinds of mineral wool. Fibrous insulation material is usually referred to as 'manmade mineral fibres', or MMMFs.

There has been a great deal of research into a possible relationship between cancer and MMMFs over the past decade or more. So far, no clear links have emerged. A couple of studies, including one in Sweden, have found slightly increased lung cancer rates in workers in factories making MMMFs, but there may well have been complicating factors, including past exposure to *asbestos* (see page 85). There was no excess in mesothelioma reported in a study of 25,000 workers in the MMMF industry in Europe, or

16,000 workers in the MMMF industry in the United States. Animal studies seem to back this up, and it is thought that the generally larger fibre sizes in MMMFs help reduce cancer risks. Further research is continuing. The work so far bodes well for mineral wool and other MMMFs not having significant long-term health implications for domestic users whose exposure is much lower.

Given the potential for irritation, it is worth being careful that any mineral wool used to insulate a water tank is not allowed to come into contact with clothes, perhaps in an airing cupboard. Most water tank insulation comes encased in plastic, to avoid contamination of water and leakage of fibres.

Polyisocyanate

This is a board used in cavity walls or the inside of buildings, non-toxic except when it burns, when it releases extremely toxic fumes of cyanide. Most material is encased in plasterboard lining, which should give some protection in a fire or at least slow down the process of combustion. There is also a slight risk of sensitization, i.e. people can become mildly allergic to the material if they handle it frequently.

Polystyrene Beads and Sheets

Used in cavity walls, on ceilings (as ceiling tiles), underfloor or on the inside of walls. There are no apparent health hazards except in the event of a fire, when they can release toxic fumes.

Polyurethane Foam

Used as foam in cavity walls, granules in roof spaces and so on. There is a slight risk of sensitization (i.e. skin irritation) and toxic fumes are released in a fire.

Urea Formaldehyde

Also known as UF Foam, or just Foam. Urea formaldehyde foam is used as a cavity wall filling, and has been the subject of intense debate because of the known and alleged health risks surrounding *formaldehyde*, which are discussed in detail elsewhere (see page 218). The US state of Massachusetts has banned the use of urea formaldehyde resin in home insulation, and Canada has forbidden future installations. Use throughout the United States has dropped dramatically, although it is still used in Britain. The British government argue that differences in house design and construction reduce occupiers' exposure in the UK.

Vermiculite
This is used as roof insulation. Apparently there is sometimes the risk of asbestiform dust being involved in vermiculite, whereupon it can cause a hazard. Studies have sometimes found asbestos dust mixed with the vermiculite and interim studies have suggested a lung cancer hazard. It is not as effective as other materials for a given thickness, and can also absorb moisture from the air, further reducing insulation properties. However, it is non-combustible.

Further Information
Good advice about all aspects of building is contained in *Hazardous Building Materials* edited by S. R. Curwell and C. G. March (E. and F. N. Spon). In addition the *Hazards Bulletin* (3 Surrey Place, Sheffield, S1 2LP) has frequently covered insulation materials, and has a number of broadsheets available.

KITCHEN HAZARDS

The Issue
Most people know that chip fires and leaking gas stoves are dangerous. But are these the only dangers? This section gives an overview of some hazards, or potential hazards, which may be lurking in the modern kitchen.

The Facts
There are a whole range of hazards, real, potential and perhaps sometimes exaggerated, which can be found in a contemporary kitchen.

Some of the most important are concerned with general levels of hygiene, as they relate to food poisoning, including *salmonella* and *listeria* (see pages 49 and 63). Others relate to accidental or intentional adulteration of food, including *pesticides* (see page 174), *nitrates* (see page 226), *additives* (see page 25), *BST* (see page 38) the potential risks of *hormones* in meat (see page 56) and new diseases such as *BSE* (see page 54). The purity of both *tap water* and *mineral water* (see page 70) have been questioned recently. The safety and effectiveness of *microwave ovens* (see page 125) has been the subject of a long-running debate. In this book we also assess the worth of some of the alternatives which claim to be healthier, including *health foods* (see page 43), *organic food*, (see page 59) and *artificial sweeteners* (see page 35).

However, it is not just food which finds its way into the average kitchen. A whole range of different cleaners, fresheners and disinfectants are likely to be crammed into the cupboard under the sink. For the sake of convenience, brief information about possible hazards from these have been collected together in the section below, although some are discussed at more length in separate sections of their own.

Most of the products in this section are not covered by legislation insisting that they are sold with labels saying what they contain, or are only controlled in some countries. In many cases it is simply not possible to say with any certainty what something will contain, and therefore impossible to make any real risk assessment.

Air Fresheners

All air fresheners which work properly have an intrinsic difficulty – they block out other smells! It may well be that the background smell would alert you to a problem, such as rotting food, leaking gas or some other pollutant. Some air fresheners are also based on *hydrocarbons* which can be dangerous to a minority of people who may experience allergic reactions. A few hydrocarbons are also potentially dangerous to everyone; for example limonene, which is used to make a lemon scent, has been shown to be a possible carcinogen in animal experiments. On the whole it is worth reducing the use of all aerosols and other air fresheners – they shouldn't be necessary if the kitchen is properly aerated and regularly cleaned.

Bleaches and Disinfectants

These are almost invariably toxic, and must be treated with care. On the other hand, letting germs build up is dangerous as well. There are a number of types of disinfectant and bleach:

- *Chlorine bleaches* such as sodium hypochlorite and calcium hypochlorite. These are highly toxic if drunk or breathed in, are powerful skin irritants and very dangerous to the eyes. They will release toxic chlorine gas if they mix with acid. Calcium hypochlorite ('Chlorinated lime') tends to release chlorine gas anyway, which can irritate eyes, skin and breathing. Use gloves and wash skin thoroughly after use;

- *Oxygen bleaches* including hydrogen peroxide and sodium per-carbonate. Similar precautions apply with chlorine bleaches, and hydrogen peroxide can also cause burns when it is in the most concentrated form, along with various degrees of lung irritation. It should never be put into a container which risks it becoming contaminated with other materials, as it can sometimes explode under pressure;
- *Quaternary ammonium compounds* are also sometimes used. These have a whole range of toxicities, from the deadly pesticide *paraquat* (which wouldn't be used as a bleach), to others which are milder.

Bleaches are dangerous and should not be used more than necessary. Too much bleach getting into a sewage works can slow down the process of breakdown and purification. There is apparently little information available on the long-term health effects.

Cling-film Wrapping

If the film used for food wrapping is made from *polyvinyl chloride* (*PVC*) it probably contains plasticizers and vinyl chloride, both of which are potentially hazardous and can migrate into food. This is especially likely if food is heated in plastic or is fatty, because the chemicals dissolve in fat. At present we have little idea how dangerous this is in practice, although there is some evidence of carcinogenicity. It is probably worth avoiding PVC cling-film for food which is still warm, for cheese and cooked meat, and for use in microwave ovens. An increasing number of cling-films are now being made from polythene, which apparently avoids these problems.

Cooking Utensils

Some pots and pans are made of *aluminium*, which is implicated in the onset of Alzheimer's Disease (see page 71). However, aluminium isn't particularly soluble in normal conditions, unless you are cooking acid foods such as rhubarb or tomatoes. If there is a history of Alzheimer's Disease in the family, it might be worth taking precautions to avoid aluminium saucepans.

Red or orange enamelled pans may contain *cadmium*. This shouldn't be a problem unless the paint is cracked or chipped, in which case it should be discarded as this makes it far easier for cadmium to leak out and into food. Cadmium is a very toxic metal. Chronic cadmium poisoning causes damage to the kidneys

and heart and can be fatal. Prolonged exposure can result in serious loss of calcium from the bones, which become brittle and
break easily. Large doses of cadmium are apparently usually
expelled fairly quickly before fatal poisoning occurs, however, and
it is exposure to dusts, vapours and chronic poisoning which are
most dangerous in practice.

Descalers
Most descalers, used for cleaning calcium deposits off kettles and
sinks, are mild acids. *Formic* and *sulphonic* acids are commonly
used on kettles. These are corrosive poisons and skin irritants and
should be kept out of the eyes. Formic acid has a highly irritating
vapour. *Citric* and *phosphoric* acids are used as 'shiners' for sinks
(and baths). Citric acid is a relatively mild skin irritant and potential allergen. Phosphoric acid is altogether more dangerous, being
highly toxic and a skin and eye irritant.

Most descalers are not strictly necessary anyway. A little calcium on the bottom of a kettle is no problem, except that it tends
to break off and finish up as dregs in the bottom of the cup.
Frequent washing out can avoid this problem. It is getting easier
to buy less toxic descalers. Check the labels.

Detergents
Environmental issues have focused on the use of phosphates in
detergents, not because of their health effects so much as their
role in freshwater pollution. However, all detergents also have the
effect of drying up the skin and can cause cracking in sensitive
people. Detergents containing enzymes ('biological detergents')
can also cause dermatitis and allergies in sensitive people. If your
skin is prone to drying or cracking, or if you are allergic to detergents, you should wear rubber gloves when washing up.

Gas Ovens
These are well known as a danger through leaks and explosions.
They can also cause death if the appliance does not burn properly
and releases *carbon monoxide*. If you feel at all unwell when using
a gas stove it is imperative that you call the gas company in *immediately*.

Oven Cleaners
The commonest oven cleaner is *sodium hydroxide*, also known as
'caustic soda'. It is a corrosive poison and an irritant, and is par-

ticularly dangerous when used as an aerosol, when there is a much greater risk of inhaling the spray. Oven cleaners are toxic enough to make it worth trying to minimize their use, by trying old remedies, such as adding salt to spilled material immediately, using baking soda as a cleaner, or simply using wire wool (although some 'self-cleaning' ovens are damaged by wire wool).

Refrigerators
There is a now well-known problem with refrigerators based on *chlorofluorocarbons* (CFCs), in that release of these is damaging the *ozone* layer. From the more immediate health and safety point of view, some refrigerators also use *ammonia* as a coolant, which can be a hazard if it leaks from the machine. This shouldn't be a problem in the normal course of events, or in a well ventilated room, but it is worth noting in case of damage to the fridge.

Stain Removers
Most stain removers are solvent-based and are thus potentially hazardous, although many brand names will not list contents on the label; take care not to inhale.

Assessment
Just when you thought it was safe to go back into the kitchen! It's as well to keep these things in perspective. The dangers of bacteria and disease in a poorly cleaned kitchen are almost certainly greater than the risks from the cleaning agents. But this doesn't mean that the hazards from the latter should be ignored. Care when using, and abandoning the more hazardous and less necessary alternatives, can reduce these risks still further.

Further Information
My handy source text for much of this section has been *C for Chemicals* by Michael Birkin and Brian Price (Green Print). *1001 Ways to Save the Planet* by Bernadette Vallely (Penguin), gives some suggestions for alternatives, as does the *Green Consumer Guide* by John Elkington and Julia Hailes (Gollancz).

LEAD

The Issue
Lead is a powerful poison which, among other things, can affect

the brain. It is cumulative, so that small amounts can build up in the body over time. Many countries have passed legislation to reduce lead levels in petrol, but it is still found in many other substances.

The Facts

Lead is a heavy metal which is malleable (i.e. soft enough to bend and shape at normal temperatures) and extremely dense. It is very toxic, especially as a neurotoxin (something which can damage the brain). The dangers of lead have been known for hundreds of years, because people who mined and refined lead had a far shorter life expectancy than their peers. The debate over the last few years has been whether lead in very small trace quantities is a perceptible hazard or not.

We can become contaminated by lead from a large number of different sources, including the following:

- *Petrol* (gasoline in North America). Lead is used as an anti-knock agent in traditional petroleum fuels for road vehicles, in the form of tetra-ethyl lead. This has resulted in the almost universal distribution of lead in all but the most remote places on earth, with very high concentrations in cities, by motorways and so on. The use of lead in petrol is a major cause of lead build-up in food plants. A few years ago the British anti-lead campaign group Clear carried out tests on lead levels in urban vegetable gardens and recommended that people did not grow food for consumption in inner city areas.

  Today, many countries are trying to reduce lead in petrol. The United States has had lead-free petrol available since 1975 and its use has been increasingly encouraged by strict emission standards. West Germany introduced severe limitations on lead levels in 1976 and is now phasing out its use altogether. The British government introduced limitations on lead in petrol in 1985, and lead-free petrol has been available in all EC countries since 1989, although the price incentives for using it vary from one country to another;

- *Paints* often contain lead, although the maximum permissible amounts have been drastically cut in many countries including, after a long struggle, Britain. The chief dangers of lead in paint today are from children eating flaking paint in old houses, or from the dust caused when stripping paint (see page 95);

- *Water pipes* have been made from lead for thousands of years.

Some commentators believe that lead poisoning was a major cause of the decline of the Roman empire, because all the top people in Rome suffered brain damage and started making the wrong decisions! There are still many lead pipes in use in Britain. A survey carried out in 1975–6 found that 7–8 per cent of water samples in England and Wales had lead levels above those set by the EC; the corresponding figure for Scotland was 34.4 per cent. Many of these lead pipes will still be in place today. The cost of making sure that all water supplies met European standards was estimated then at between $1,500 million and $5,000 million; prices today would be far higher;

- *Lead solders* in food cans result in contamination of food. These are being phased out in some countries, but can still be found in many imported foods;
- *Lead in building* is used for many purposes, but is best known as a roofing material. This is probably one of the safer uses, because the lead is not released as particle or dust in normal usage, although there are health risks for people who frequently install (or steal!) lead from roofs.

Lead also enters the environment in many minor ways. *Lead shot* used by anglers has poisoned swans in many areas and probably adds to lead levels in fish which are caught and eaten. Lead used in shotgun pellets can cause problems if they are left inside animals killed for the pot. Lead is also used in trace amounts in many metal alloys.

The health effects of lead are well known. Acute lead poisoning results in stomach pains, tremors, headaches, irritability and, in severe cases, coma and death. However, the effects of more chronic lead levels in children have been the subject of considerable controversy for the past twenty years. Lead is known to affect the nerves and brain at quite low levels and many authorities now accept that levels of lead commonly found in urban environments are having a small but significant effect on the mental functioning of children who live there. On the other hand, a number of studies have not found any significant difference. There have also been some cases where experts remain divided about the significance of results, notably a large piece of research around 'Spaghetti Junction', the large and famous motorway interchange in Birmingham, England.

One of the problems with studies of this kind is that we have few if any accurate ways of measuring intelligence. Some of the best known, including the infamous '11–plus' in Britain, have since been shown to be wholly inaccurate. On the whole, the balance of opinion is now that low levels of lead *are* probably dangerous, but we still have a long way to go before this can be quantified at all accurately.

Minimizing Risk

- Certainly it is worth changing to lead-free petrol if your car is capable of using this. (Most modern cars will accept lead-free, although some small cars such as the citroen 2CV and the older minis, cannot);
- If you go out picking blackberries, or other *wild food* (see page 78), do not collect these next to a busy road, car park or roundabout where lead is likely to have accumulated;
- Think very carefully before growing vegetables right next to a busy road. If you have a large enough garden, it might be worth making the vegetable garden as far from cars as possible;
- Check whether water pipes in your house are of lead. This will only be the case in an old house. If so, consider changing them. At the very least, run the water for some time before using for drinking, to give lead which has settled in the standing water a chance to flush away. Some *water filters* (see page 75) are quite good at removing lead from water and are worth installing if you have lead pipes;
- Check with the local water compnay if there are any lengths of lead pipe in the system leading to your house. If so, it might be worth getting some tests of average lead levels, although you may have to pay for this service.

Further Information

There were a lot of books published about lead a few years ago, although many are now out of print. Try *The Lead Scandal* by Des Wilson (Heinemann); Des Wilson was chairman of Clear, the campaign group influential in forcing the British government to reduce lead levels in petrol. A detailed analysis of health effects, which is out of print but worth getting from a library, is *Lead and Health*, published by the Conservation Society in Britain.

Useful groups include Citizen Action in the UK (3 Endsleigh Street, London WC1) which incorporated the Clear campaign.

LEGIONNAIRE'S DISEASE

The Issue
Legionnaire's Disease seems to be increasing. Is there anything we can do about it?

The Facts
Legionnaire's Disease is a form of pneumonia, caught by inhaling water droplets containing *Legionella* bacteria. About 200 cases a year occur in Britain, and these often include some fatalities.

Legionella bacteria also cause a non-pneumonic illness rather like flu, called 'Pontiac Fever' which has occurred most commonly in the United States. It is named after an outbreak of fever in an office in Pontiac, Michigan, in 1968. The cause was unknown at the time but examination of preserved tissues from laboratory animals affected by the fever strongly suggest that *Legionella* was the cause.

Legionella bacteria are very common in water and also sometimes in soil. Given the right conditions, they can multiply extremely quickly. In Britain, the authorities estimate that the maximum concentration of *Legionella* for normal safety is 10,000 bacteria in each millilitre of water. This means that bacterial concentrations at this level won't generally do you much harm. When conditions allow them to multiply more quickly, far higher levels can build up. During an outbreak centred on the British Broadcasting Corporation's London headquarters in 1988, concentrations of up to 10 million bacteria per millilitre were measured, i.e. a thousand times above the safety limit.

Legionella bacteria only become dangerous when they are inhaled, and for this to happen they need to be present in small droplets of water mixed with air. Water vapour contaminated with bacteria capable of causing Legionnaire's Disease is only usually found in artificial conditions. The commonest sources of outbreaks are water cooling systems. This is partly because *Legionella* only flourish in certain water conditions and the relatively warm water in cooling systems helps them to multiply, and partly because the water systems frequently release vapour containing bacteria into the atmosphere. Once this happens, even people some way from the source can be affected.

The bacteria are extremely common in water systems. A recent survey by the Public Health Laboratory Service in Britain found *Legionella* present in 70 per cent of hospital hot and cold water systems and 67 per cent of hotels. It must be stressed that in the vast majority of cases the bacterial levels were too low to pose a health risk, but they do all have the potential to cause an outbreak unless contained.

Cooling systems are not the *only* cause of infection. Others include the water which remains in shower heads in domestic bathrooms. There has been some speculation that in these cases bacteria get into the body through the eyes or mouth rather than being inhaled, but we still don't know for certain.

Most people are apparently highly resistant to Legionnaire's Disease. Susceptible people include the elderly, especially if they already have respiratory problems; smokers; and people taking drugs which reduce their natural defences. However, no one is totally immune, and once water vapour containing high levels of bacteria is released into the atmosphere people of all ages and fitness can be affected in an outbreak.

For example, in the BBC outbreak mentioned above, contaminated water vapour drifted out from the Broadcasting House headquarters in central London and eventually affected at least 86 people, including three who died. Many of the victims had never set foot inside the BBC building, and contamination occurred throughout much of the West End, extending from Oxford Street to Regent's Park. Cases like these are not infrequently reported from around the world.

Assessment
Risk level after exposure is low for young and healthy people. However, risks should be zero, as explained below.

Minimizing Risk
Professor Donald Acheson, Chief Medical Officer for the Department of Health in Britain, has stated that Legionnaire's Disease is wholly avoidable so long as people in charge of standing water tanks, such as those used in heating systems, take a few simple precautions.

Unfortunately, most of the people affected by Legionnaire's Disease don't have control over the water systems which harbour the disease. For those who do have some influence, Legionnaire's Disease can be prevented by:

- Adding a simple chlorine substance into standing, non-drinking water tanks every few months to kill the bacteria. Older and more complicated systems should be treated more regularly;

- Maintaining high levels of hygiene in water systems, to prevent build-up of organic matter which can serve as food for the bacteria;
- Eliminating areas from the water system where water can stand for long periods e.g. 'dead leg' areas of reduced flow, kinks in the pipe, sharp corners on water tanks and so on;
- Anyone involved in the design of a new building, or water system, should ensure that it is built in such a way that areas of potential build-up of *Legionella* are minimized. This includes constructing pipework in the ways described above, making water towers capable of being cleaned and avoiding any standing water;
- In addition, anyone with a domestic shower should make sure that it is bacteria free. This can be done by frequent cleaning, or by running hot water (about 50°C) through for a few minutes before showering. Self-draining showerheads are now available to avoid this problem.

What Governments Could Do
Most countries don't have laws in place to control Legionnaire's Disease. In Britain the Health and Safety Executive have issued guidelines similar to those outlined above; although these don't have force of law, failure to comply with them would be regarded as grounds for prosecution.

One way of ensuring that proper controls took place would be a licensing system for all water supplies in which *Legionella* is likely to breed. This would be similar to Britain's annual MOT inspection system for cars. Institutions would have to submit their systems for an initial inspection, then pay for annual inspection and be subject to random tests, to ensure that sufficient was being done to reduce the risk of the disease.

Further Information
The World Health Organization have produced a useful report on risks of Legionnaire's Disease and its control, *Environmental Aspects of Control of Legionellosis: Report of a WHO Meeting*, available from the WHO office in Copenhagen, Denmark. The American Industrial Hygiene Association produces *Recommendations and Work Practice Guidelines for Ventilation and other Work Practice Standards* which includes Legionnaire's

Disease. The Department of Health in Britain can also offer advice.

LICE AND NITS

The Issue
Everyone can get head lice, nits (the eggs of the lice) and body lice sometimes. Peoples' first instinct is to reach for the lice shampoo: but the pesticides used in these preparations are extremely hazardous, and can be absorbed through the skin and enter the body.

The Facts
Getting lice and nits isn't a sign of being dirty. Quite the reverse. Lice very sensibly tend to choose a clean head of hair to set up house. Some people are naturally prone to picking up external parasites, while others almost never do. There is a general rule (by no means always true) that blond people are less likely to attract head lice than dark-haired people, whatever their skin colour. Children in school are quite likely to pick them up, as one infected child can spread the problem through a whole class in a matter of days.

The pesticides used in preparations designed to kill lice are amongst the most toxic known, capable of being absorbed through the skin, and potentially damaging to health. Among those which you may find (and once again there are differences between individual countries as to what is allowed) are:

Carbaryl: used against head and body lice. Carbaryl is a carbamate pesticide which is poisonous to humans and can be absorbed through the skin. The US *Farm Chemicals Handbook* recommends avoiding contact with the mouth, eyes and skin. There is a lot of evidence of carcinogenicity, including a detailed monograph from the International Agency for Research into Cancer. Mutagenicity is also suspected, and the US Environmental Protection Agency suggest that there is evidence of teratogenicity. It is an anti-cholinesterase compound. US experiments listed by the National Institute for Occupational Safety and Health (NIOSH) suggest possible neurological and reproductive effects on rats.

Lindane: (also known as *Gamma HCH*) used against head and body lice. A persistent organochlorine insecticide. Poisonous to humans and can be absorbed through the skin, which is an important route of entry. Notes issued by the Department of Health and Social Security (now the Department of Health) in London suggest that it can also cause damage to the central nervous system. It is a suspected carcinogen and teratogen. It has been banned or severely restricted in over fifteen countries. (See also *timber treatment* page 245).

Malathion: used against head and body lice. An organophosphorus compound which is poisonous to humans and can be absorbed through the skin. The US *Farm Chemicals Handbook* also lists it as harmful if it comes into contact with the skin. It has been identified as a potential teratogen by the International Labour Organization of the United Nations and is an anticholinesterase compound. The World Health Organization have cited evidence for emotional effects on people working with malathion.

Assessment
It is worth remembering that actual concentrations of these chemicals are fairly low and you shouldn't be exposed to them very often. Nonetheless, these are dangerous poisons which can be absorbed through the skin, so using them in shampoos is worth avoiding if at all possible. The section below lists some alternatives and precautions.

Minimizing Risk
Even if you have lice or nits, there are a number of less toxic options open to you:

- The nit comb. For head lice, combing the hair thoroughly with a fine comb (specially bought for the purpose) can often get rid of nits and lice, especially if the hair is washed in a mildly medicated shampoo afterwards, perhaps with some medicinal disinfectant added. This is obviously much easier with short hair;
- If you do use a special shampoo, malathon is probably the best of a bad bunch. Read the instructions carefully and ensure that it doesn't get into either mouth or, especially, the eyes;
- Minimize contact with the body. Don't use shampoo on the

head while bathing, as this can expose the whole of the body to the pesticide (and at a time when pores are open because of the warm water). Use cool water and wash the hair separately in a bowl or basin.

MICROWAVE OVENS

The Issue
Microwave ovens are being used in ways which allow bacteria on food to survive. There are also persistent fears about the radiation effects from old or faulty microwave systems.

The Facts
Microwave ovens do not 'cook' food in the conventional way. They bombard food with microwaves, as the name suggests, which agitate molecules of food and water, and it is this agitation which causes the food to become hot. This allows food to be cooked, or reheated, very quickly.

Microwave ovens are used in places serving food where there is a need for rapid turnover in a small space: these include certain types of fast food restaurant; catering departments on trains, aeroplanes and ferries; shops selling take-away food with an option on having it warm; a growing number of cafes and restaurants; and, increasingly, people in their own homes. At the moment, the United States apparently leads the world in domestic microwave ovens, with at least one in 70 per cent of homes. People are increasingly using microwaves for almost any kind of cooking.

When eating out, you can usually tell if something is cooked in a microwave if it comes quickly, is extremely hot, yet is in a cold bowl or plate. Microwaved soups often have a tell-tale dried ring around the top. And, of course, you can often hear the ping of a microwave in the back of the shop as it finishes its cycle.

Microwave ovens have a number of undoubted advantages. They do speed up the cooking process considerably and, if used properly, can produce very tasty food. (People disagree about whether food from a microwave tastes as good as food cooked slowly in a conventional oven.) Microwaves also use considerably less energy than conventional cookers, which is cheaper and environmentally beneficial.

However, there are a number of problems as well:

- Microwaves are notorious for having 'cold spots' where the food is not cooked. This is because the waves are not uniform and miss out areas, especially in older models or where the oven is not set correctly. A rotating platform helps overcome cold spots to some extent, although problems can still arise;
- Agitation of molecules, and hence heating, only usually occurs in the top 2 centimetres of the food at most. Deeper heat penetration is by conduction, as in a normal oven. However, due to the quick turnover in the microwave the temperature in the centre of larger items of food is unlikely to be high enough to kill bacteria;
- Both *listeria* and *salmonella* bacteria can survive microwave cooking;
- Thawing food from a freezer in a microwave is similarly uncertain and can result in frozen spots remaining;
- The speed and apparent efficiency of the microwave can tempt people to use it in inappropriate circumstances. For example, microwave ovens are frequently used to reheat food which has been standing for too long and has built up high bacteria levels; in these circumstances microwaves do not sterilize the food;
- Some of the older microwaves have been linked to a number of disorders in people who use them constantly, including damage to the eyes.

Dr Richard Lacey, the well-known authority on bacteria in food, has written: 'Microwave ovens have a use in the home, but it is limited.'

Minimizing Risk
If you have a microwave, only use it for the things it is best suited to cook safely. These include:

- Baked potatoes;
- Soups made from 'safe' ingredients (i.e. not meats or other materials likely to have high bacterial counts,) such as dairy products;
- Reheating *recently* cooked food;
- Heating sauces;
- Heating drinks.

Further Information
Information on the safety of microwaves can be found in *Safe*

Shopping, Safe Cooking, Safe Eating, by Richard Lacey, (Penguin).

OFFICE POLLUTION

The Issue
Modern offices include many materials which are known or suspected of being damaging to health. Many of these problems are fairly minor, others are more important. This section gives a brief overview of the main issues.

The Facts
The office environment has attracted a lot of attention from the green movement recently, both in terms of the environmental impact of office work (such as use of recycled paper, minimizing energy use and so on) and because of the toxic effects of various materials and processes used in the modern office. Concern about the latter has led to the theory of *sick building syndrome* (see page 145) and the appointment of special facility managers to improve conditions for office workers.

These issues are increasingly being recognized as important by national and international agencies. The International Labour Organization (part of the United Nations) has written that office employees 'run the risk of exposure to harmful products ... they often do not realize the danger'. Lack of understanding about risks has been identified in official reports for years without making much impression, either in terms of removing hazardous substances or stimulating better staff training.

Some of the major issues are covered elsewhere in the book. They include the increasing use of desk-top computers with *visual display units* (see page 129); choice of *building materials* (see page 95) and paint (see page 135); the risks of *Legionnaire's Disease* (see page 120) and associated problems with *air conditioning* and *ventilation* (see page 84); and the use of *fluorescent lighting* (see page 103). Some offices still have problems with *asbestos* (see page 85), and *passive smoking* (see page 68) continues to be contentious issue throughout the world.

Other concerns centre on specific materials, machines or practices in offices. Literally thousands of office materials can have health effects, ranging from skin complaints (probably the commonest problem of all) to development of cancer. In this section there is a brief overview of some of the main problems not covered elsewhere.

Aerosols: range of possible toxic materials released, see page 83.

Adhesives: most glues and adhesives include *solvents* which are discussed at length on page 236. Effects of inhaling too much can include dizziness, headaches, bronchial symptoms, eye, skin and throat irritation and some are also carcinogenic.

Carbonless copy paper: research in both Sweden and Britain has identified risks of dermatitis when handling large quantities.

Carbon paper: dermatitis has been reported by workers at the Post Office in Britain.

Cleaning fluids (including *type cleaners*): risks of dermatitis.

Inks: some inks are poisonous, others cause dermatitis.

Photocopiers: operators are exposed to a range of chemicals, many in gaseous form, which vary depending on the type of machine in use. They can include *ozone* (see page 157), *toners* and *carbon monoxide* (in dry copiers) and *ammonium* (in wet copiers). Some also emit *nitrogen oxides* and *selenium*. Many of these gases are fairly toxic and users should avoid breathing them in continuously.

Typing correctors: many are *solvent*-based, with a wide range of possible health effects. Solvents used include 1,1,1 trichloroethylene (which is also an ozone depleter), titanium resin, and toluene.

Waxes and polishes: risks of dermatitis.

Assessment
Don't despair! The risks of most of these are minor irritants rather than deadly poisons. Dermatitis, which is by far the commonest complaint listed above, is only likely to be an issue if someone is handling large quantities of material, or has especially sensitive skin. However, there *are* enough problems to make it worth trying to minimize risks. And the *overall* effect of a whole

range of toxic material combined with poor lighting and ventilation, bad office furniture and mental stress can indeed add up to real problems.

Minimizing Risks

- Keep tops on glues, correcting fluids and other solvent-based liquids when not in use. Solvents are very volatile, so they can build up quickly in a confined space if left exposed to the air. If your correcting fluids keep drying up it means you are letting too much dangerous vapour into the air;
- Don't site photocopiers in enclosed spaces, and make sure that they are adequately ventilated;
- Ensure that offices are well ventilated, and that air conditioning does not simply recycle stale air, thus keeping pollutants in the office;
- The office manager should keep note of possible risks, and ensure that staff who show signs of particular problems (e.g. skin complaints, constant runny nose, sore eyes and so on) have their work rescheduled to avoid the materials thought to be causing the problem;
- There should also be adequate staff training about potential risks.

Further Information

There are some good books available on office environment, which include pollution issues. In Britain look out for *Office Workers Survival Handbook* by Marianne Craig, published by the Women, Work and Hazards Group, and distributed by Hazards (P O Box 199, Sheffield S1 1FQ). In the United States *Office Work Can be Dangerous to Your Health!* by Jean Stellman and Mary Sue Henifin (Fawcett Crest) gives a good breakdown of issues and many useful sources and addresses. See also the National Academy of Sciences Report *Indoor Pollutants*, (NAS, Washington DC, 1981).

Visual Display Units

The Issue

Computer users have probably heard rumours about possible health problems for pregnant women. Are there any grounds for these suspicions? And are computers bad for your eyes, back or wrists?

The Facts

Desk computers with a TV screen are called visual display units, or VDUs, in Britain, and visual display terminals (VDTs) in the United States. There are now millions of them in use all over the world; in Britain the latest figure of 2 million professional VDU operators in employment is probably an underestimate.

Because of the speed with which VDUs have been introduced into the workplace, many employers only have a very rough idea about how they should be used, or about possible health effects. There has been quite a lot of debate in the popular press, in medical journals and through the trade union movement. Some European countries have already introduced legislation controlling the amount of time people can work on VDUs in any one day.

Health implications are complicated to assess. Some health effects are already quite clearly understood; they are well defined, physical and psychological results of sitting in front of a keyboard for long periods of time. Other, more controversial, issues involve the possible impact of radiation from the computer and the effects on eyesight of looking at a screen for hours on end. These are still unproven.

These are six main areas of concern:

- Problems caused by sitting at a keyboard for long periods of time;
- Eye strain and injury;
- Skin problems;
- Hearing loss;
- Long-term impacts of radiation;
- Effects on reproduction.

Problems Caused by Sitting at a Keyboard

On the face of it, this might appear to be no different to using a conventional typewriter. To some extent this is true; any job which involves sitting in a fixed position for hours can lead to posture problems, and some of the publicity given to VDUs in this respect comes from our greater awareness of it as an issue today. But VDUs differ from conventional typewriters as well. The keyboards tend to be in different positions. It is possible to type many more characters per minute, which increases the risk of physical damage known as *repetitive strain injury* (RSI). And operators face more repetitive typing, without the breaks for putting in more paper, or, on a manual system, hitting the carriage return.

The result is that people (usually women) who use word processors continually, day after day, often suffer from a range of aches and pains and, sometimes, from more serious muscular injuries as a result. A bewildering array of possible health problems have been identified, including tenosynovitis, bursitis, carpal tunnel syndrome, Dupuytren's contracture, cervicobrachial disorders and muscle strain. A survey by the Civil and Public Service Association in Bristol, England, found 64 per cent of VDU operators suffered pain or stiffness in their shoulders and necks, almost half had wrist trouble and 31 per cent suffered finger problems. These health effects are very serious and widespread, but can be tackled by a few simple changes in both operator practice and the arrangement of the workplace.

Skin Problems

Some of the plastics used on machines, the cleaners and inking units can cause skin problems in sensitive people. The whole issue of skin complaints from office materials is dealt with in more detail in the section on page 129.

Hearing Problems

VDU operators are only likely to suffer hearing problems if they are working in a small space surrounded by lots of noisy printers. Unfortunately, some people have to do exactly that.

Eye Strain and Injury

This is much more contentious. Studies over the past decade have shown the VDU operators are likely to suffer increased eyestrain, but the reasons for this are still subject to fairly impassioned debate. For example, the Swedish Board of Occupational Safety and Health found 77 per cent of VDU operators suffering from eyestrain. The Canadian Labour Congress found 87 per cent of full-time or frequent VDU operators suffering from eyestrain compared to 64 per cent of non-VDU clerical operators. A similar comparison by the National Institute of Occupational Safety and Health (NIOSH) in the United States found 91 per cent of heavy VDU users with eyestrain compared to 60 per cent of clerical workers. A US survey found that eyestrain was greatest among those involved with repetitive input work, and a British study found that eyestrain increased with the amount of time spent on the computer.

However, several other detailed studies have concluded that there is no direct link between VDU use and damage to sight. Some opticians believe that VDU use simply exposes previous eye weaknesses, or that problems come mainly from misuse. The National Academy of Sciences in the United States published a report which was very critical of studies linking eyestrain with VDUs, although one researcher insisted that his dissenting opinion be added at the end of the report.

At the moment, the debate about eyestrain rests mainly on the legibility of the screen; i.e. that a screen is simply more difficult to read than a piece of paper. However, there is also a growing debate about the role of radiation from the screen in causing cataracts. NIOSH in the United States has examined the evidence and dismissed these claims, while in Britain the Health and Safety Executive is apparently still keeping an open mind on the matter. And whatever the experts say, it is undoubtedly true that the majority of VDU operators *believe* that they suffer greater eyestrain than their colleagues who don't use computer terminals.

Radiation and Reproductive Effects

The controversy hots up even more when the possibility of radiation from the VDUs causing health damage is examined. There have been a large number of studies throughout the industrialized world which have looked at the health effects of VDU work on pregnant women – on their success in having children, and on the subsequent health of the children. Apparent correlations between VDU work and health problems have been found in studies in Japan, the United States, Canada, Denmark, Britain and other countries. Ricardo Edstrom, medical director of the Swedish National Board of Occupational Safety and Health is on record as saying that: 'We can no longer rule out the possibility that radiation (from VDUs) could affect foetuses.'

However, this is still a long way from saying that there is definitely a problem. Most of the studies have been small-scale and not statistically significant. Some may well have been started because several women in an office or factory suffered birth problems at the same time, thus rousing suspicions but also unbalancing the statistics. No long-term epidemiological studies have been attempted, and no studies of the impact on fathers carried out.

One possibility suggested to explain increased birth problems amongst VDU operators is that the pulsed extremely low frequency (ELF) radiation given off from VDU screens is more dangerous than much stronger non-pulsed radiation fields. Some experiments carried out in Sweden and Poland suggest that pulsed ELF radiation can increase foetal mortality, but other studies elsewhere suggest that there is no correlation. In research sponsored (but not published) by IBM, Professor Arthur Guy of the University of Washington, Seattle, wrote:

> Though it is highly unlikely that there is any relationship between birth defect clusters and VDU emissions, the clinical work ... does indicate that there could be a relationship.

This argument is going to run and run. If you are pregnant and anxious, take heart that a Scottish industrial tribunal determined that a pregnant librarian had been unfairly dismissed after refusing VDU work and that her fears 'were by no means ill-founded'.

Assessment
Very difficult to make at the moment. Certainly, people using VDUs should watch out for back problems, eyestrain and other physical injuries (and also for mental strain resulting from sitting at a keyboard for too long of course). But the questions about birth defects are still very much up in the air. Given the enormous amounts of money being made by computer firms, and the way in which all industrialized economies are now totally reliant on computers, getting a straight answer to these problems in the next few years may not be easy.

Minimizing Risk
Many of these suggestions are simple, already quite well known, and ignored by employers and VDU users almost everywhere!

- Get set up properly. With the computer on a table, at the correct height for typing and seeing the screen, on a comfortable chair that supports your back and with everything positioned in a way that doesn't throw any part of your body under stress. If you suffer from aches and pains you're doing something wrong!
- Get proper advice about *how* to sit at the computer. Many VDU users make their problems worse by slumping over the

keyboard, leaning too near the screen and so on. Your local trade union or your employer should be able to give advice;

- Make sure the screen is as clear as possible. This means using a good screen, positioned so that it is not obscured by reflected light, and at the right distance for you to be able to read it without straining. There are also a number of filters which can be fitted onto the front of screen and thus ease eyestrain;
- Some of the psychological strain can be eased by reducing noise. This means positioning noisy printers away from people, or fitting an acoustic hood (this problem is getting easier as more people have laser printers or other quiet machines);
- Neck and back strain can be reduced by using document holders when copying something, so that you don't continually have to lean over sideways to read text;
- However well the machine is filtered and set up, the only way to avoid or reduce eye and muscle strain is to keep a check on the amount of time you work on the VDU in any one day. The Norwegian government has now put a four hour limit on VDU use in offices, and recommends that people take a fifteen minute break every hour to give eyes and backs time to recuperate. It is important to ask employers not to demand unreasonable periods in front of the VDU screen, through the trade union representative if necessary. As an employer, try to organize work to ensure that people are not so stretched that they cannot take adequate breaks. And if you are self-employed, keeping a check on your own hours – always the most difficult of disciplines!

What Governments Could Do

It is sometimes hard to remember that ten years ago virtually no one had a VDU. For example, in Britain none of the major environmental groups had computers until around 1983-5. Yet now they are in use everywhere, with very few controls over hours worked, adequate screening and so on. And governments are often reluctant to get involved in these issues. Adequate and cautious working practices are urgently needed in most countries, but they'll only come if enough people demand them.

For a start, there should be realistic limits on the number of hours that someone can be forced to work on a VDU in any one day. Secondly, screens should meet set standards for clarity and protection. And thirdly, there should be an international committee set up to assess the effects, if any, of radiation from computers, with a large enough epidemiological study to answer the questions one way or the other.

Further Information

There are some good books available on VDUs. In Britain, try the *VDU Hazards Handbook* by Ursula Huws, published by the London Hazards Centre (Headland House, Grays Inn Road, London WC1X 8DS). In the United States, *Office Work can be Dangerous to Your Health* by Jeanne Stellman and Mary Sue Henifin, (Fawcett Crest) contains information about VDTs and, for more detail, see *Terminal Shock – The Health Hazards of VDTs* by Bob De Matteo (NC Press Ltd, 1985) and *VDT Health and Safety: Issues and Solutions* by Elizabeth A. Scalet (Ergosyst Associated, 1987). There is also a bimonthly newsletter in America: *VDT News: The VDT Health and Safety Report* (PO Box 1799, Grand Central Station, New York NY 10163).

PAINTING AND DECORATING

The Issue

Paint contains a range of hazardous additives. Some of these can be breathed in during and after painting, while others are a risk to children who eat flaking paint, or during paint stripping. Other materials used during decorating are also hazardous to health.

The Facts

Paints

There are a number of possible health hazards from using paints:

- Inhaling solvents in the paints which can be dangerous to health;
- Ingesting (i.e. breathing in or eating) paints which contain high levels of hazardous additives, especially lead;
- Exposure to pesticides (especially fungicides) used in paint to stop mould growing;
- Fire risks from oil-based paints.

The hazards of *solvents* are discussed in detail on page 236. A number of hazardous solvents are added to paints, including *white spirit, xylene* and *trichloroethylene*. Spray paints can give off particularly high levels because the solvents are very volatile and can evaporate quickly from the spray. A number of the solvents become additionally toxic if inhaled through a cigarette, and some are also irritating to the skin. A few are also very flammable. The levels of solvents are likely to be fairly high for as long as there is a strong paint smell in the freshly painted area.

In the past, very high levels of *lead* (see page 116) were added to paints. Some countries still permit lead in paints above the levels recommended by international bodies. For example, in Britain it was only in 1986 that a maximum lead concentration of 600 ppm (parts per million) was set, although this had previously been adopted in the United States, and had been recommended by the UK's Royal Commission on Environmental Pollution.

Now that most countries have controlled lead in paints, the risks are mainly from old paints, which are still in place. And remember, paints don't have to be that old to contain lead. A survey by the compaign group Clear in Britain in 1983–4 found over 80 per cent of paints sold had lead above the 600 ppm level, and some had levels more than ten times higher.

People can get contaminated from lead in existing paint-work in two ways. Some young children have a habit of picking and eating flakes of paint. This is part of the phenomenon called 'pica', which means the tendency for young children to eat non-food substances. Studies in Glasgow, in Scotland, some years ago found that children eating paint in old houses could receive high doses of lead as a result.

However, the main danger from lead today is from stripping old paint. People renovating houses over fifty years old or so can easily encounter paint with very high lead levels a few layers down on walls and woodwork. Dry sanding to remove lead is particularly hazardous in that it releases paint as a dust which can be breathed in. Burning or other stripping methods can create lead-rich fumes which can be toxic to painters and decorators over a long period. Use of chemical paint strippers is another option, but these contain their own hazards if used in confined spaces. Once the dust is in the house, it can remain for along period of time and continue to poison inhabitants by getting into food, onto cooking surfaces and so on.

Some old paints also contain high levels of *chromium, chromium (VI)* or *chromate*. This is highly toxic if inhaled in the same way as lead, and can cause a range of effects, including dermatitis, ulcerations and cancer. In modern paints, *cadmium* is often added to yellow paints, and *titanium oxide* to white paints. Both of these should be treated with caution.

Another possible health risk is from the *fungicides* which are added to paint to prevent mould. Some of these can be quite toxic; for example *tributyl tin* and *carbamates* are used. These should be treated as pesticides rather than paints.

Decorating Materials
There are also problems with some of the materials which are used alongside paints.

Paintbrush cleaners and restorers are made from organic solvents, especially *naptha* and *dichloromethane*. These both have harmful vapours and are skin irritants (especially dichloromethane), including severe irritation to the eyes. All paintbrush cleaners are likely to be flammable.

Paint strippers come in two main types. *Caustic alkali* strippers can cause serious skin burns. You should avoid breathing in vapour and wear gloves and eye protection. *Solvent* paint strippers include dichloromethane often used with *methanol*; or *trichloroethylene* if being used to remove varnish. It is worth trying to avoid using these in a confined space.

Wallpaper pastes are usually cellulose based. These should present no problem unless they have fungicides added (which should be marked on the container) in which case they should not be handled.

Assessment
Risks from lead in old paint are real and should be taken very seriously. Exposure to solvents is potentially dangerous as well, both in the short term in making you feel drowsy or unwell, and in terms of risking more serious, long-term effects.

Minimizing Risk
There are serious grounds for avoiding stripping old paint if you suspect it has high lead levels. Curwell and March write in their book *Hazardous Building Materials – a Guide to the Selection of Alternatives:*

Wholesale removal is generally neither feasible nor cost effective, and if not done with scrupulous care, will probably make matters worse by releasing lead dust and particles around the home to be inhaled and ingested. When paintwork is sound, and children are not exposed to it, it is unlikely to present any serious hazard, whatever its lead content.

They recommend leaving it in place or painting over it.

However, some people will want to or have to strip old paint, and, unfortunately, we all have to get around to putting on new paint every so often. The following should help minimize the risks involved in this process:

- Remove all loose paint exposed to young children;
- Always wear a mask when stripping *any* paint;
- If you have to strip paint over 50 years old, use chemical paint stripper to avoid dust or fumes;
- When using stripper
  —thoroughly ventilate the room
  —leave vacant until the smell has gone
  —wear gloves and a mask
  —keep away from flame, heat or plastic;
- Check whether fungicides have been added to the paint.

Further Information
S. R. Curwell and C. G. March (editors) *Hazardous Building Materials – a Guide to the Selection of Alternatives* (E. F. Spon) gives a detailed breakdown of hazards and how to avoid them.

RADON IN BUILDINGS

The Issue
A couple of years ago the British National Radiological Protection Board dropped a bombshell by saying that 1,500 people a year could be dying of lung cancer in Britain because of natural radon gas seeping into their homes. Is this just a way of diverting attention from the problems of the nuclear industry, as some people have suggested, or is it really an important issue worth worrying about?

The Facts

Most rocks contain some uranium. As this decays, it releases radioactive *radon gas*. This in turn decomposes into other radioactive isotopes, known as *'radon daughters'*, which form tiny dust particles in the air. Once inhaled, radon daughters can become attached to the lungs and cause the development of cancer, many years in the future.

Problems occur when radon reaches high levels in buildings. It used to be thought that this only occurred when there was an insufficient number of air changes per hour, as a result of draught-proofing and insulation. When this happened, it was argued, radon emanating from bricks and other building materials could reach dangerously high levels. This idea stemmed from work in Sweden, where stringent energy conservation measures were introduced after the 1973 oil crisis, and high radon levels were later measured in buildings.

Although you'll still often hear it stated as hard fact that insulation increases radon exposure, experts studying the problem have long abandoned this theory. Quite the reverse; it is now believed that radon enters buildings through tiny cracks in walls and floors. The British National Radiological Protection Board (NRPB) believe that radon is sucked into buildings because the pressure is slightly lower indoors than in the surrounding soil. Internal pressure is reduced by the effects of wind forces on walls and windows, and across chimneys, and is reinforced by higher air temperatures in the winter. (One exception is if warm air is blown into a building as part of the heating system, as this raises pressure and forces air out again.) The NRPB identified the following routes of entry for radon:

- Cracks in concrete floors;
- Air gaps between construction joints;
- Cracks in walls;
- Cracks in suspended timber floors;
- Gaps around services (electricity, water, gas);
- Through cavities;
- By percolating up through stone walls.

The possible role of insulation in radon build-up and the effects of radioactivity from building materials have not been carefully evaluated, but are no longer thought to be major causes. Improved ventilation may make the problem worse, according to the NRPB, because it decreases pressure inside the building even more, sucking in greater quantities of radon.

A further argument against the insulation theory is that concentrations of radon vary dramatically from place to place. They are likely to be highest in granite areas (which probably helps explain some of the Swedish problems). Granite is now well-established as a source of radon gas. In Britain, some of the highest radon levels are found in parts of Cornwall and Devon. Other areas with relatively high radon levels include: Cumbria, Snowdonia, Anglesey, a wide band running along the Grampian mountains between Aberdeen and Fort William, the north-eastern tip of Scotland including the Orkneys, parts of Dumfries and Galloway, and some areas in Somerset. A few houses with higher than average radon levels have also been found in Derbyshire and Yorkshire. In badly affected buildings, radon levels can be thousands of times higher than in the outside air.

It is important to stress that radon probably only causes a real problem when it is at exceptionally high levels, and we still don't quite know why a particular house has high radon levels. One of the things which started the radon scare in the United States was when a worker at a nuclear power plant, one Stanley J. Watras, triggered off the radiation alarms at work because of the radon lingering on his clothes from home. The concentration of radon measured in his home is said to be the highest ever recorded, reaching 100,000 becquerels per cubic metre.

Health Risks

As stated in the introductory section, it is impossible to tell what has caused any particular cancer. Radon has been identified as an important cause of lung cancer through epidemiological studies of radon miners, including black miners in Namibia and Indian miners in parts of the United States. (It is usually only the poorest people who end up down radon mines.) Epidemiological studies have also shown that people with high radon levels in their houses suffer higher than average incidences of lung cancer.

Careful studies over a number of years suggest that radon could be a major cause of lung cancer in countries where it is likely to occur. In March 1988, the NRPB increased its estimates of possible death-rate from radon exposure by 250 per cent on previous figures, which shot it up into the largest single source after smoking, accounting for 5–10 per cent of all lung cancers in the industrialized world.

Mike O'Riordon, department head of the Radiological Measuring Department, reported that: 'Lifetime risks imply that 1,500 people may die of lung cancer each year in the UK from indoor exposure to radon and radon daughters.'

These figures are broadly similar to those worked out by the Environmental Protection Agency in the United States. The EPA estimates that a lifetime exposure at 150 becquerels per cubic metre gives a risk of lung cancer ranging from 2.4 to 9 per cent; using these figures they calculate that 5,000 to 20,000 of the 130,000 lung cancer deaths per year in America could be caused by radon.

Figures like these are even more alarming when it is considered that only a relatively small proportion of houses are at serious risk from radon exposure. The NRPB estimates that over 20,000 British families are exposed to radiation doses above the limits proposed for workers in the nuclear industry as a result of radon gas.

However, these figures have been attacked by the British Institution of Environmental Health Officers who believe that they are still too low. The IEHO estimated that at least 60,000 families were exposed to radiation levels above the 400 becquerels per cubic metre of air suggested as a 'safety limit' by the government. They consider these limits to be too high in any case, and recommend that they should be halved. If this proposal was adopted, 300,000 British families would be officially 'at risk'.

Indeed, some of the statistics being bandied around are very alarming. The British government has set a national radon limit of 20 millisiverts a year; the NRPB estimates that someone exposed to this level constantly runs a 5 per cent lifetime risk of developing lung cancer. According to the magazine *Building Design*, the current Health and Safety Executive limits in Britain give someone constantly exposed to these levels a 1 in 8 chance of dying of lung cancer.

However, as you will probably have realized by now, nothing is every quite so clear cut in the world of environmental health. There have also been a number of studies which dispute those of the NRPB, and of other national and international bodies reporting on radon. For example, in late 1988 an American study looking at the incidence of lung cancer in women living in high radon areas throughout the world found slightly *lower* lung cancer levels than the average.

However, this study did not control properly for smoking, so that the results may well be distorted. Certainly, no one is claiming that all the data for calculating risk levels are complete, and some of the figures are probably alarmist. A large epidemiological study in south-west England, being carried out by the Imperial Cancer Research Institute, should provide some kind of overview when it is completed.

Assessment

At the moment, the balance of evidence suggests that radon is a real problem, although the scale of the problem will continue to be debated for years. Using the most pessimistic figures, anyone living in a house with very high radon levels will be at quite high levels of risk, and would certainly be wise to take precautions to prevent radon from entering the building.

Minimizing Risk

Various guidelines have been drawn up for controlling radon. This section is based on those drawn up by the Building Research Establishment in Britain.

- Find out if you have a problem. If you are in an area known to have high average radon levels, ask for a test. In Britain, contact the Building Research Establishment, Garston, Watford WD2 7JR. *Don't* assume that you have high radon pollution just because you live in one of the areas listed earlier, or identified in the press, because a lot depends on very local conditions, the state of the house, and so on;
- If you have a radon problem, take professional advice about how to cut down radon entering the building. The BRE guidelines centre on stopping radon seeping from the soil into buildings, either by improving barriers or by diverting the gas from underneath the building. Costs for this range from £100 for sealing floor cracks, to £1,000 for installing an underfloor extractor, and £10,000 or more for installing a new, impermeable floor;
- Researchers in a number of countries, including Sweden and the United States, are looking at various methods for removing radon gas from within buildings. It appears that every building is likely to be different and, before rushing into panic measures, it is worth getting professional advice (and measuring radon levels again once you've finished to check that the preventative methods have worked!).

What Governments Could Do

Radon has been ignored as an issue for a long time. There is still a great deal of evaluating to be done, and an urgent need for concise, clear advice for people living in radon-rich areas. A programme of distributing information to all householders at risk, or potentially at risk, is needed in many countries, along with much clearer guidelines about what is and isn't safe.

Further Information

A clear introduction to the whole issue of health effects from radiation is *Radiation Risks: An Evaluation*, by David Sumner (Tarragon Press, Glasgow). In addition, many governments have now produced basic information about risks, and the World Health Organization has published a report: *Indoor Air Quality: Radon and Formaldehyde: Report of a WHO Meeting*.

SCHOOL LABORATORIES

The Issue

The place where many people have their closest contact with dangerous chemicals, radioactive particles and toxic fumes is in the school laboratory. Safety standards in some schools leave a lot to be desired.

The Facts

Once upon a time, chemistry teachers had one of the shortest average life expectancies in Britain. School experiments were carried out with minimal safety requirements; highly volatile and carcinogenic liquids were passed around to sniff; and many chronic poisons, such as mercury, were routinely handled. Safety has improved a lot since then, at least in theory, but there are still a number of potential risks involved.

In Britain, the 1974 *Health and Safety at Work Act* introduced tighter controls over the operation of school laboratories, although many of the chemistry teachers will have been trained and started working before this was introduced.

There are big differences in legal exposure limits to chemicals in school laboratories between countries. Standards set by the Health and Safety Executive in Britain are sometimes lower than those of the National Institute of Occupational Safety and Health (NIOSH) in the United States, or the American Conference of

Governmental Industrial Hygienists (ACGIH). Swedish standards tend to be even more stringent.

There are a number of possible risks in school laboratories. Some things to look out for are:

- Low ceilings and inadequate ventilation;
- Too few fume cupboards, or fume cupboards set in the wrong place so there is limited visibility of experiments;
- Lack of an earthing system for the electrical circuits;
- Water taps set next to electrical sockets;
- Very small sinks, so that the area around is always wet;
- No extractor fans;
- Very large class sizes so that supervision is difficult;
- Classes left unsupervised;
- Reactive chemicals stored together in the same place;
- Poor labelling of chemicals (either names or hazard warnings);
- Poor standards in disposal of chemicals;
- Poor fire safety and a lack of fireproof cupboards for chemical storage;
- Re-use of plastic petri dishes for work in biology laboratories;
- Poor disposal of animal remains in biology laboratories;
- Inadequate safety equipment, including goggles, gloves, pipette fillers (used instead of sucking chemicals up by mouth, as used to be the case).

Old schools, with laboratories built decades ago, are particularly likely to have outdated safety precautions because the whole room may well have been constructed in ways which wouldn't be permitted today.

Assessment
Most schools are well aware of the dangers and take adequate precautions. But not all do, and if you are worried about conditions in your school laboratory, it is certainly worth taking the matter up with the head teacher.

Further Information
There are a number of good reference texts about safety. A broadsheet available from the *Hazards Bulletin* (PO Box 148, Sheffield, S1 1FB) gives a useful summary of risks and comparisons of safety exposure limits for chemicals in different countries. *Safety in Science Laboratories* from the British Department of

Education and Science is quite old (1973) but still worth hunting out for basic information. The main teachers' union in Britain, the National Union of Teachers, has a *Safety Representatives Handbook*, (available from NUT, Hamilton House, Mabledon Place, London WC1H 9BH).

There are also many sources of information about maximum permitted exposures limits to different chemicals. In America the ACGIH Threshold Limit Values are published annually (ACGIH, 6500 Glenway Avenue, Building D-7, Cincinnati, Ohio 45211-4438, USA). The *NIOSH Recommendations for Occupational Safety and Health Standards* (NIOSH, CDC, Atlanta, Georgia 30333, USA) are also available. In Sweden, the *Translation by the Swedish Plastics and Chemical Suppliers Association of Ordinance AFS 1984-5* gives the Swedish figures for maximum chemical exposures in English (National Swedish Environmental Protection Board, S-171 84, Solna, Sweden).

SICK BUILDING SYNDROME

The Issue
Experts now believe that people working in offices who feel continually unwell aren't always just neurotic or shirking, but that bad management plus a whole range of pollutants can result in something which has become known as 'sick building syndrome' or SBS.

The Facts
This book won't attempt a learned discourse on SBS. However, it is a useful concept which ties together a number of the issues described on other pages.

Sick Building Syndrome describes a complex mix of environmental stresses on office workers. Symptoms include headaches, stress, eyestrain, minor infections and a general 'unwell' feeling. Until recently it was dismissed as slightly hysterical imaginings. Now, SBS is increasingly recognized as a real problem, and 'facility managers' are being employed to tackle it, and thus increase productivity.

A survey by the Buildings Use Studies group in London found that 80 per cent of office workers in the sample experienced symptoms of ill health which they associated with the workplace. The commonest symptoms were lethargy (57 per cent), stuffy nose (47 per cent), dry throat (46 per cent) dry or itching eyes (46 per cent) and headaches (43 per cent). Clerical and secretarial workers are most likely to suffer from SBS. This was explained by a number of factors, including the more sedentary working patterns as compared to managers and the different conditions in offices for people further down the scale of an organization. Buildings with air conditioning (see page 84) consistently had higher rates of SBS than those with outside ventilation and, in Britain, public sector buildings appear to have higher rates of SBS than private sector, perhaps because they are generally older and not so well maintained. Lack of control over the office environment was cited as a factor in SBS by many people in the survey. There has also been a recent study in Sweden which linked SBS to reduced worker productivity.

Sick Building Syndrome has many causes, including poor ventilation, dirty air (i.e. air which is being constantly recirculated, or just polluted air), incorrect temperature settings, and minor pollutants (see page 127). There are also a number of psychological factors which can get involved; these are none the less serious in their effects on human health. Lack of a view, inability to control temperature by opening a window, noise levels and lack of privacy can all help make people ill. Boring repetitive work, and particularly work on *VDUs*, can make the problem worse.

Health Effects
Can be very real for minor complaints. There are also a whole range of long-term health problems associated with constant stress or exposure to chronic levels of pollutants. SBS should be taken more seriously than it is by most employers.

Minimizing Risk

- A combination of all the techniques outlined under office pollution, Legionnaire's Disease, VDUs, air conditioning and so on;
- If you're in a big company which can afford it, try asking for a facility manager, perhaps through the trade union representative.

Further Information
Three are now some detailed studies available on sick building syndrome. In Britain, Building Use Studies Ltd produce *The*

Office Environment Survey (Sheena Wilson and Alan Hedge, Building Use Studies, 14/16 Stephenson Way, London NW1 2HD) which looked at the incidence of SBS in a range of office buildings. For a detailed technical brief, see *The Facility Manager in Today's Office* published by Maclaren in Croydon, UK. There is also an Association of Facility Managers (30 Abbots Road, New Barnett, Herts) and a Facility Manager Magazine.

In the United States, some help can be obtained from the Center for Occupational Hazards (5 Beekman Street, New York NY 10038; telephone (212)-566-1699).

PART THREE:
OUTDOOR HAZARDS

AIR POLLUTION

The Issue

Air pollution is acknowledged as being one of the worst environmental problems today. Made up of a vast range of different pollutants, and international in effects, air pollution has been implicated in turning rivers acid, killing trees and even altering the earth's climate. But despite an enormous amount of research, there is still considerable disagreement among scientists about its direct effects on our own health.

The Facts

Until quite recently, there has been a mistaken belief in the rich countries that air quality was improving. Early legislation which cut down visible pollution (mainly smoke) transformed city centres, especially in Western Europe. Blackened buildings were cleaned for the first time in a hundred years and, at least in North West Europe, the old 'pea soup' fogs became largely a thing of the past. In the United States, despite many attempts at pollution control, heavy smogs still occur in big cities at many times of the year.

In the 1980s, new problems started to come to light: *acid rain* from sulphur and nitrogen oxide pollution, which killed the fish in lakes in huge areas of Europe and North America; *tree dieback* from a mysterious combination of air pollutants, acidification and drought, which destroyed huge areas of the German Black Forest, along with trees in many Eastern European countries, Britain and parts of North America: the *ozone* hole, created from over-use of aerosol cans and fire extinguishers (see page 157), and the spectre of the *greenhouse effect* and changing climate, as a result of carbon dioxide from burning fossil fuels. Air pollution moved back to the top of the political agenda again.

This section doesn't look at the effects of individual pollutants, which are discussed in their own sections. Instead, it tries to give an overall feel for how important air pollution is to health, how things are changing, and what, if anything, people can do themselves to cut down risks.

In order to understand the possible effects on health, it is important to know how atmospheric pollution is changing, and what pollutants we can expect to encounter. Today the world can be divided into three broad 'chunks' for the purposes of discussing air pollution:

- *Western industrialized countries,* including West Europe, North America and Australasia. Here, some kinds of pollution have been substantially reduced. But not as much as the politicians would like us to believe, and at the same time, new pollutants have come to take their place;
- *Eastern European countries:* despite the partial liberalization of governments in the east, pollution problems have not changed as yet. Here poor to non-existent pollution laws, combined with heavy industry and high-sulphur coal, have produced the worst air pollution in the world;
- *The South:* in the Third World, air pollution is currently increasing rapidly, as industrialization continues apace. Again, there are insufficient regulations and often major social and economic blocks in making laws work in practice. Western firms have frequently moved South where they can get away with less stringent pollution control than is possible at home.

An overview is important, because air pollution, more than any other environmental issue, doesn't stop at national borders. And when problems such as the *ozone hole* or global warming are considered, it becomes a truly international issue.

The bulk of air pollution still comes from the industrialized countries, where, despite a lot of hype about pollution control, there continue to be major problems. Attempts at controlling pollution by dispersal through building high chimney stacks haven't got rid of the problem but have simply spread it out over a wider area. Air pollution now affects more of the planet than ever before, and is more variable. Inner cities and industrial areas still suffer the worst problems, but remote areas also receive occasional bursts of pollution because of chance occurrences in the weather. Although a few pollutants have declined, including smoke and sulphur dioxide in some countries, others have increased, or appeared for the first time, in the last few years. Whereas in the past there were areas of very high pollution, with other places remaining fairly pristine, today the situation is altogether more complex. We haven't begun to understand the full implications of this for health and ecology.

There are literally thousands of pollutants to be found in an urban area, but a few are particularly important. *Sulphur dioxide* and various *nitrogen oxides* are still extremely common. *Ozone* can be formed in some conditions, especially in the summer. There are a whole range of factory effluents and *vehicle pollutants* (see page 202), including *carbon monoxide* and those from *diesel engines* (see page 199). *Dusts* (see page 103) are common throughout urban areas. There is also a host of relatively minor pollutants (that is, minor once you get out into the open air), including *solvents* (see page 236), the effects of *passive smoking* (see page 68), bacteria such as those causing *Legionnaire's Disease* (see page 120), pollution from *dry cleaning* (see page 100), *bonfires* (see page 168), and so on, and so on.

In some cities, especially where the air is still and sunny during the summer, thick *smogs* can still occur. In Europe, a number of southern cities are especially badly affected; Athens is notorious for its bad air quality. Many of the Eastern European cities are badly affected, including Cracow and Wroclaw in Poland and Prague in Czechoslovakia. In North America, Los Angeles, San Francisco and other west coast cities are best known for their smog, but effects in Boston, New York and the industrial cities of the north are probably just as bad. Smogs intensify air pollution problems, both by lowering people's resistance and because they hold pollutants in the air and increase the 'dose' to humans.

There is still an enormous debate about how great a risk this air pollution actually poses to human health. There is little doubt that extreme air pollution is a problem, such as that occurring in smogs. The Great Smog of London in 1952 killed an estimated two thousand people in a fortnight and finally caused a large enough public outrage to force through the introduction of the first *Clean Air Act*. Smogs in the United States are viewed with enough alarm that people are advised to stay indoors on particularly bad days. These come quite frequently; in 1989 in Boston there were over twenty days when people were advised to stay indoors.

Smogs cause immediate problems such as bronchitis, breathing difficulties, chest pains, a general feeling of being run down and reduced resistance to infection. The very young, very old and people with existing respiratory problems or heart disease are most at risk. These facts are established, and should be taken seriously.

However, the role of air pollution in creating longer-term health problems, and especially cancer, is much more contentious. Some specialists believe that air pollution only directly accounts for a tiny proportion of total cancers. Others put the figure much higher.

One of the problems with any assessment is that the most likely cancer to be associated with air pollution, on the face of it, is lung cancer. Here, smoking is such an overwhelmingly important factor that it tends to obscure other factors. However, there is a growing body of opinion which says that in some cases smoking and air pollution work together in promoting lung cancer, i.e. someone who smokes is more likely to develop cancer from air pollution. This distinction might seem a bit academic, but it is important.

In the United States, there have been some attempts to put figures to these speculations. A large study which looked at cancers in 25 states, and over a million people, concluded that at least 11 per cent, and more likely 21 per cent, of all lung cancers were due to air pollution.

A symposium organized by the US Department of Health held in Stockholm, Sweden in 1977 concluded that:

> Combustion products of fossil fuels in ambient air, probably acting together with cigarette smoke, have been responsible for cases of lung cancer in large urban areas, the numbers produced being in the order of 5–10 per 100,000 males per year.

A follow-up symposium, in 1983, reaffirmed these basic findings and suggested a dual link between smoking and air pollution.

Hopefully, these figures should decrease. Despite the generally gloomy prognosis about pollution, there have been large decreases in some of the most dangerous urban air pollutants over the past few years, including polycyclic aromatic hydrocarbons and smoke. New legislation, especially in Europe, should cut out many pollutants from vehicles. On the other hand, we are only just starting to look at the implications of many of the new toxic materials now escaping into the atmosphere, from electronics factories, new solvents, continually increasing traffic and so on.

Assessment
The probability is very high that air pollution is currently killing people in cities. We know far less about general pollution levels in the countryside. Hopefully, steps have been taken to improve the situation, but the effects of these won't be seen immediately. There is still much more to be done.

Minimizing Risks
Unfortunately, there's not a great deal people can do about

general air pollution except stay out of the most polluted areas if they are especially susceptible. People who are likely to suffer most in smogs and other heavily polluted atmospheres include:

- People already suffering from respiratory diseases such as bronchitis, emphysema and asthma;
- Those with heart complaints or angina;
- Very young infants and unborn children;
- Anyone with reduced resistance to infection through previous illness or problems with the immune system;
- Those with specific allergies triggered by air pollution.

In addition, it is best not to exercise too hard, or in high temperatures, during periods of peak pollution. Specific advice is found in the section on jogging and cycling (see page 198).

Further Information
Some American cities have regular recorded phone messages about smog, and a few also have live phone lines for detailed information. Smog warnings are broadcast on local television and radio.

There are few accessible books giving an overview of air pollution and health, but *Air Pollution and Health* by Claire Holman, prepared for Friends of the Earth in Britain, gives a good overview and a summary of the main references.

BRACKEN

The Issue
Bracken is a very common fern growing throughout Britain and much of Europe. It is also a powerful *carcinogen*, and there are fears about the effects of exposure on humans through breathing in bracken spores and drinking milk and water from bracken-covered areas.

The Facts
They are still fairly sketchy. Bracken is almost certainly a carcinogen, causing stomach cancer, but the significance of different routes of exposure, and the general levels of risks to humans, are still not very well understood.

There are a number of different sources, or possible sources, of carcinogens in bracken. The spores are carcinogenic in mice, and may well be in humans as well. There are also at least three carcinogenic substances which can be leached from the fronds and rhizomes (which function a bit like roots) by water. Possible risks which have been identified include:

- Breathing in spores of bracken by walking through colonies of the fern during the late summer and early autumn, when spores are released;
- Drinking water which has drained off bracken-covered slopes, or is contaminated with spores;
- Eating food which has been grown using bracken as mulch;
- Eating young bracken fronds;
- Drinking milk from cattle which have been contaminated with bracken, either through eating or by drinking water containing soluble carcinogens from bracken.

It is also dangerous to use bracken as a bedding for animals in situations where they are likely to eat it. Bracken can also be acutely dangerous for them, causing poisoning, and sometimes death, through haemorrhaging.

The evidence for levels of risk remains very incomplete. Research in Japan, where young bracken fronds are considered a delicacy, shows that people who eat them daily have triple the chance of developing oesophageal cancer.

Evidence of risk from walking through bracken is much less certain, because detailed studies have not been carried out. The sporing time is late summer and early autumn, and one bracken frond can produce up to 300 million spores. However, bracken does not set spores very regularly, and normally spreads through vegetative reproduction via its tough and persistent rhizomes. Spores may well only be produced when bracken is 'stressed' in some way, i.e. by bad weather conditions or disease. Most walkers, or people working on bracken-covered slopes, willl probably not be exposed to bracken spores.

Bracken in water is a more complex problem. There are possible risks from spores in the water and from the soluble carcinogens. A study carried out by the Welsh Water Authority concluded that: 'bracken contamination of large water supplies is not a prime source of stomach cancer in North Wales' due to dilution by other water. The WWA believe that spores are largely eliminated or inactivated by the treatment process and that most float so that they do not get taken into the water system. However, there are apparently no studies of the effect of bracken carcinogens in water drawn from individual wells and small reservoirs, where levels could be much higher.

There is also concern about carcinogens from bracken turning up in milk supplies. The Welsh Water Authority study quoted above concluded that this was a possible cause of the high levels of gastric cancer which occur in North Wales. Other researchers still believe that water-borne carcinogens could play an important role, and some studies suggest that stomach cancer is higher in areas where water drains through bracken-covered catchments.

Minimizing Risk

Despite the ambiguous nature of the evidence, it is probably worth avoiding the most obvious sources of bracken, by:

- Not using bracken as a compost or mulch;
- Avoiding pushing your way through bracken during periods when it is releasing spores (which will appear as a fine dust as you brush the vegetation). This doesn't happen very often but when it does it will be in the late summer or early autumn;
- People drinking milk or water from areas of high bracken cover may consider buying milk from further away, or using an alternative water source. Alternatively, if bracken is covering the catchment of a small reservoir or well, it can be eliminated, or at least substantially reduced, by cutting twice a year for three years. The first cut should be when the fronds are just a few inches high, and the bracken should be broken off rather than cut cleanly, so that the plant is damaged as much as possible and the sap 'bleeds'. (And, of course, never cut at the times of year when bracken is sporing!)

Further Information

A good source of information about all aspects of bracken is *Bracken in Wales* by the Senior Technical Officers Group, Wales, published by the Nature Conservancy Council in February 1988.

OZONE

The Issue

People have a right to feel confused about ozone. One minute we hear that ozone is the thing that makes going to the seaside so healthy, then ozone pollution is being blamed for tree death and crop damage, and is considered to be an important component of office pollution and smogs. Finally, *lack* of ozone in the stratosphere is said to be a major health threat in places.

The Facts

Ozone can be a health and environmental problem when it occurs too near the earth's surface. However, the thin layer of ozone in the upper atmosphere (the stratosphere) has an essential role in keeping out dangerous types of solar radiation.

The ozone in the stratosphere is a layer of gas, mostly occurring 20–25 kilometres above the ground. It occurs at very low concentrations; if all the ozone in the stratosphere were brought to ground level and evened out it would only be about 3 millimetres thick. However, it plays a vital role in cutting down certain wavelengths of the sun's radiation – including ultra-violet radiation – which are dangerous to humans and other living creatures.

Normally the amount of ozone in the stratosphere remains about the same. Some is broken down, and more is created through chemical reactions, usually above the equator. From there, it spreads out all over the earth's surface. However, in recent years we have been releasing increasing quantities of chemicals which can break down ozone, and thus destroy the protective ozone layer.

The chemicals which deplete ozone include:

- *Chlorofluorocarbons* (CFCs) which are found in aerosols, blown foam such as that used to wrap up burgers, and many refrigerators;
- *Halons* from some types of fire extinguishers;
- Some *solvents* such as *methyl chloroform* and *carbon tetrachloride*;
- Pollution from *supersonic aircraft*;
- To a much lesser extent, a whole range of minor chemicals.

Any chemically-stable compound containing chlorine or bromine can help destroy the ozone layer. Some of these gases also contribute to the greenhouse effect, which is the major climatic change which may be occurring as a result of industrial pollution.

At the moment, the 'ozone holes' are temporary, generally forming in the winter and disappearing during the summer, but this may change unless the ozone-destroying chemicals are drastically reduced. The most famous hole is that over the Antarctic, which has been measured as the size of the United States and as deep as Mount Everest. Scientists think that a similar hole may be forming over the Arctic and small, local, holes have been found elsewhere.

These ozone holes are already having effects on people's health. In Australia, which is affected by the ozone hole over the Antarctic, levels of certain sorts of skin cancer are increasing dramatically, including some varieties of fatal melanoma. It has been calculated that as little as a one per cent depletion of the ozone layer will lead to an extra 70,000 cases of skin cancer worldwide, every year. In addition, ultra-violet radiation can cause eye damage even at quite low concentrations, and reduces the efficiency of the body's immune system, making people more susceptible to other diseases.

Nearer the ground, ozone is an environmental pollutant. It is strongly suspected of playing a role, and sometimes a very major role, in the widespread forest dieback being found in Europe, North America and, increasingly, in other areas as well. This *Waldsterben* (which is the German word now accepted internationally for the phenomenon) is partially caused by high ozone levels. This ozone is formed from reactions between various industrial pollutants in the presence of sunlight. Two of the most important precursors of ozone are *nitrogen oxides* and *hydrocarbons*, both of which are especially plentiful from vehicle exhausts.

Despite the hoary old myths about the health-enhancing effects of ozone near the sea, ozone at high levels is now regarded as an industrial pollutant. It can impair lung function and make breathing more difficult, lower resistance to disease, and is a possible mutagen.

It is not at all clear how dangerous ozone is to human health at current pollution levels. It has certainly been suspected of causing problems when it is emitted from *photocopiers* in confined spaces (see page 128). However, ozone can be a more general pollutant when it is a component of the smogs which frequently occur in parts of North America and in southern European cities. Some areas of California issue regular health bulletins about smogs; the South Coast Air Quality Management District, for example, lists the following health problems associated with ozone in smog:

• Coughing, wheezing, chest tightness or pain, dry throat, headache or nausea;

- Shortness of breath or pain during deep breaths;
- Lung damage;
- Reduced resistance to infection;
- Tired feeling;
- Impaired athletic performance.

People particularly at risk are identified as very young, elderly, people with reduced resistance to diseases, including lung diseases, and those who exercise hard or in high temperatures during peak ozone periods.

Assessment
The effects of the ozone hole are already being experienced in some places, and are likely to get worse. The impact of ozone on human health when it occurs near ground level is far less well known, and will probably take some time to quantify.

Minimizing Risk
The ozone hole must be countered in two ways: reducing personal impact on the ozone layer in the first place, and taking suitable precautions against getting skin damage if you are in high risk areas.

You can make a personal contribution to cutting down ozone destruction by doing the following:

- Avoiding *aerosols* containing CFCs (which probably means *any* without a note saying they are CFC free);
- Not buying food wrapped in blown foam cartons, and telling people in shops and fast food chains why you won't accept them;
- Making sure that any fire extinguishers you have do not contain halons; the largest fire extinguisher firm in Britain have now said that halons are never essential;
- Avoiding any solvents which contain methyl chloroform (trichlorethane), or carbon tetrachloride, which are both important ozone depleters. (Carbon tetrachloride has already been banned for safety reasons in many countries);
- Avoiding refrigerators which use CFCs in the cooling system, and take professional advice about getting rid of old fridges to ensure that trapped CFCs are not released into the atmosphere;
- Not using methyl bromide as a fumigant, as this is also an

ozone-depleting chemical (as well as being extremely toxic).

Advice about avoiding skin problems is given under *sunbathing* (see page 166), and avoiding ozone problems at work under *office pollution* (see page 127).

What Governments Could Do
Most governments throughout the world are still being overly complacent about the greenhouse effect. Although industry has reacted fast in removing CFCs from many aerosols, and elsewhere, levels are still not falling fast enough, and several official reports have been exceptionally gloomy about the consequences. Far more radical approaches to reducing the ozone-depleting chemicals are urgently needed.

Further Information
There are a great many books being written about the ozone layer at the moment. One of the best for the general reader, although now slightly out of date, is *The Hole in the Sky* by John Gribbin (Corgi Books). Friends of the Earth and Greenpeace worldwide have produced many publications, practical guides and handbooks on ozone depletion and how to fight it in your own home. FoE in Britain have reports available about aerosols, CFCs in buildings, in wrapping and so on. See also *The Use of CFCs in Buildings*, by Curwell, Fox and March (Fearnshaw 1990).

In the United States, a good introduction (although again considerably outdated in legislative aspects) is *The Sky Is The Limit: Strategies for Protecting the Ozone Layer* by Alan S. Miller and Irving M. Mintzer (World Resources Institute 1986). A detailed, gloomy, overview is given by the EPA paper *Future Concentrations of Stratospheric Chlorine and Bromine* by John Hoffman and Michael Gibbs (EPA 400-1-88/005, 1988, Office of Air and Radiation, Washington DC).

PCBS

The Issue
Polychlorinated biphenyls (PCBs) were a widely used class of industrial compounds, which have been gradually withdrawn over the past twenty years because they were found to be toxic, persistent, and capable of being concentrated to very dangerous

levels in the food chain. Unfortunately, large amounts of PCBs are still scattered in the environment, and attempts at disposing of them safely are causing international problems.

The Facts

There are several different types of PCB. Although toxicity varies, and is not always very well known, all PCBs are toxic to some degree or another. Most are likely to be carcinogenic, and some are also acutely poisonous. They give off toxic fumes of *dioxin* when burnt at temperatures below 1200°C.

Manufacture of PCBs has virtually ceased. They have only been used on closed systems, such as electrical transformers, for years. But there are still large quantities to be found in the environment and in predators at the top of food chains such as birds of prey and seals. Levels in some marine habitats, including the North Sea around eastern Britain, are causing particular concern.

Although PCBs are known to be extremely toxic, it has been difficult to pinpoint precisely what ill-effects they can have on marine life, humans and other animals. PCBs were one of substances cited by marine scientists as contributing to the weakening of seals off the British coast during 1989, when a mysterious virus wiped out whole populations. One prominent marine scientist said recently that there were enough PCBs left in the environment to cause the extinction of virtually all types of marine mammal. While most researchers would remain more optimistic about the survival of at least some whales and seals, it is universally admitted that polychlorinated biphenyls are among the most dangerous industrial waste pollutants known. PCBs are concentrated in land food chains as well. Like *dioxins* they are found in human mother's breast milk, in eggshells and elsewhere in the environment.

The practice of 'dumping' hazardous wastes in some African, and other Third World countries has added to the problem. During 1988 and 1989 a number of scandals brought to light the practice of some waste firms from the North paid to dispose of hazardous waste safely, and doing so by dumping it in certain West African countries, where conditions for safe disposal do not exist. Both the people and the environment are put into direct contact with some of the most toxic wastes known, including PCBs. The ship Karin B became particularly infamous when a large number of countries, including Britain, refused her entry because of the waste being carried.

In the rich countries, the greatest risks from PCBs probably come from mismanagement in *waste treatment plants* (see page 243). At the moment, many people believe incineration is the safest way to dispose of PCBs, the alternatives being dumping (which doesn't dispose of them at all) and storage until safer disposal methods are developed. All have their risks. Greenpeace are opposed to incineration and favour storage on the surface until safer disposal methods are found, but this risks catastrophic leaks if there are accidents or sabotage in the stores. Incineration only works if it is carried out correctly at very high temperatures (and there will usually be small leaks of dioxins in any case). Some of the waste disposal firms have built up bad reputations for illegal or sloppy operation of the plants. Choosing a disposal method inevitably becomes a choice between two evils.

Most people are only likely to come directly into contact with PCBs if they have been used in old capacitors in *fluorescent lights* (see page 103).

Minimizing Risk

There isn't much we can do ourselves about general levels of PCBs in the environment, beyond lobbying for better disposal methods. Refer to the separate listing on *fluorescent lights* for safety instructions about their handling and disposal (see page 103).

Further Information

The World Health Organization has produced several important publications about PCBs, although their conclusions are fairly conservative. See, for example, *PCBs, PCDDs and PCDFs in Breast Milk: Assessment of Health Risks* by Grandjean *et al* (WHO); and *PCBs, PCDDs and PCDFs: Prevention and Control of Accidental Exposure* by J. H. Rantamer *et al*, (WHO).

SEAWATER

The Issue

Seawater around the coasts of industrial countries is often full of sewage, industrial pollutants and other toxic materials; are there real risks to health?

The Facts

In 1959, the British Public Health Laboratory Service reported that: 'Bathing in sewage polluted water carries only a negligible risk to health, even on beaches which are aesthetically very unsatisfactory.' I leave it to your imagination what that means!

Until recently, it has been assumed that the seas around most of the coasts of Europe, North America and Australasia were fairly pristine, apart from purely localized pollution problems caused by sewage outfalls or industry. It has long been accepted that very polluted water is dangerous to health. The practice of bathing in the Ganges in India is recognized as a public health disaster. Nearer to home, the waters of some Mediterranean resorts had become so polluted as to be unsafe for bathing; people lay on the beach and dipped in freshwater swimming pools. But apart from that, the oceans were considered to be large enough to dilute and neutralize most of what we could throw into it.

Two things have helped change this attitude. The first is that series of surveys which have shown unequivocally that coastal waters are polluted over much of the industrialized world, especially with sewage. A European Community survey, following an EC Directive on bathing water quality in 1985, found high levels of pollution throughout the Community. Only about half of Britain's designated beaches, for example, met EC mandatory water quality criteria. Britain has been threatened with legal action by the European Community if beaches are not improved.

An independent survey carried out by staff at Surrey University found that, according to their data, only just over a third of Britain's beaches were clean enough to satisfy EC regulations. A continent-wide survey by the Bureau Européen des Unions de Consommateurs (BEUC, a consumer organization) found pollution levels so high that they described the situation as *'une veritable catastrophe'*.

The second major change is that governments have started accepting that swimming around in sewage-polluted water is probably unhealthy. This has long been the attitude in North America, where microbial standards have been related directly to incidence of disease amongst bathers. Standards in both the United States and Canada are about five times tighter than those operating in Europe.

In Britain, there has been a gradual shifting of the establishment position. A survey carried out by a professional consumer survey group for Greenpeace UK found a statistically significant increase in reported stomach upsets in people swimming in polluted water compared with those on a similar, but unpolluted, beach. At about the same time, a spokesman for the British Medical Association admitted that: 'The BMA's policy statement of 1959 that there is no danger to health from swimming in sewage polluted water may well be obsolete ... consequently we would wish to re-examine the problem'.

Part of the reason for continuing uncertainty is that we don't always have a very clear idea of what pathogens are actually in the water. It is thought that most gastro-enteritis attacks in seawater are caused by viral pathogens, while most tests look for bacteria instead – because this is cheaper. However, the most popular indicator bacteria, *faecal coliform*, probably disappears faster than viruses, which may distort the surveys and suggest that the sea is cleaner than it really is.

Assessment

It looks as if there are real risks of getting gastro-enteritis if you swim in contaminated water; i.e. a tummy upset and diarrhoea. Risks of more serious diseases, including typhoid, cannot be ignored, but are probably fairly remote.

Minimizing Risk

- Use your eyes and nose! If the water looks dirty, i.e. you can see sewage or you are obviously near a sewage outfall, don't swim;
- Ask someone local. I know one Welsh resort where all the townspeople avoid a particular beach near a short-sea sewage outfall, but unsuspecting tourists bathe there every year;
- You can also ask local tourist boards which beaches meet EC standards while planning your holiday;
- Look out for local information. In some European countries (France for example), local authorities have to display information about seawater pollution fairly prominently. In Britain, a few beaches which are clean and fairly litter free have been awarded a 'blue flag' by the Blue Flag Campaign UK, which is identifying clean beaches for tourists;
- Most people who report getting gastro-enteritis attacks after swimming have immersed their head in water. If you're a bit worried about the water, don't duck your head underneath the surface, and try not to get any in your mouth.

What Governments Could Do

There is still a great deal of time and money needed on stopping pollution of coastal waters. Banning the dumping of raw sewage into the sea would be a good start. As an interim measure, all governments could follow the lead of those who provide proper, accurate information to allow people to make up their own minds about whether they want to swim or not.

Further Information
See *The Good Beach Guide* from the Marine Conservation Society (4 Gloucester Road, Ross-on-Wye, Herefordshire, HR9 5BU) for information on 180 good beaches in Britain.

SUNBATHING

The Issue
Sunbathing is one of the most popular pastimes of holidaymakers in affluent countries, and having a 'good tan' is an essential part of style and attractiveness for both sexes. Most of us know it is not good in excess, but how dangerous is sunbathing really?

The Facts
A certain amount of exposure to the sun is good – some people would say essential – to health well-being. The sun's energy gives us vitamin D and lying in the sun makes many people feel very relaxed, which is in itself good for the health.

However, over-exposure to the sun can lead to a number of problems, some minor and others very serious. One almost inevitable result for people who overindulge in sunbathing, or simply work in the sun, is premature ageing of the skin, which dries up and cracks. Today, we can see the first generation of real sixties sun-worshippers well into their forties and fifties, and a fairly dried up bunch some of them look as well, despite lavish applications of softening creams and skin toners!

Much more serious is the risk of a range of skin cancers. These risks are *increasing*, partly due to the greater amount of time that people spend in the sun, and partly because of the effects of holes in the *ozone* layer which mean that more of the dangerous radiation in sunlight reaches the planet's surface (see page 157).

Degree of risk depends to a large extent on skin colour. Pale-skinned people will never go deep brown, however long they lie in the sun, and people with sensitive skin are also at greater risk of skin cancers. People of northern descent usually have a high risk of developing skin cancers, and there is even a risk scale known as 'celticity'. Rates are especially high in Scandinavia and Germany, high in Britain and Ireland, Australia, and the USA, low in Mediterranean countries and in any areas where people have dark skins.

Assessment
This is a real risk, often ignored because 'everyone sunbathes'.

Minimizing Risk
For people determined to sunbathe, the following will help minimize risk:

- Start slowly. As little as two minutes each side on the first day. Try never to go pink; this is a sign that the skin is damaged;
- Use a good sunblocker, suitable for your skin colour and with a moisturizer to prevent the skin from drying out. All creams need to be re-applied regularly if sweating or swimming;
- Watch out for skin growth which could be cancerous, and consult your doctor immediately if you find one; moles or spots are particularly likely to develop cancerous tendencies – if they become larger and a deeper colour you should immediately consult a doctor. Most skin cancers *can* be successfully treated if caught early enough.

SUN BEDS

The Issue
Many people use sun beds to help give themselves a year-round tan, brown their skin before going on holiday, or just because they think it is good for them. It isn't.

The Facts
A certain amount of natural sunlight and sunbathing is, indeed, good for you. It encourages the formation of invaluable vitamin D and is therapeutic. But too much is bad, especially in those parts of the world where the *ozone* layer has been depleted through air pollution.

Many people living in climates with less sunshine have taken to using sun beds, in health clubs, beauty parlours and their own homes. There is a lot of fashion pressure for white people to have a deep tan, and pale skin is supposed to look unattractive. (The reverse was true a hundred years ago, when only the rich could afford to stay in the shade and everyone else had to work outside!)

However, use of 'artificial sunlight' – which usually includes large amounts of ultra-violet (UV) radiation – has a number of known human health effects, including:

- Drying out and premature ageing of the skin;
- The promotion of a variety of skin diseases, including lethal and non-lethal skin cancers;
- Damage to the eyes, unless special darkened goggles are worn.

In 1985, the American Medical Association wrote that:

> There is no known medical benefit to be obtained from cosmetic taning. Exposure to high intensity UVA radiation in the tanning booths currently in vogue is a health hazard.

You can't get much plainer than that, but many people are still unaware of the dangers. In April 1990, a woman wrote to Dr Miriam Stoppard, a well-known medical 'agony aunt' in a British weekly magazine. She asked how she could reverse the drying out of her skin that she had noticed after heavy use of the sun lamp all winter. Dr Stoppard replied, quite rightly, that she couldn't – such damage is irreversible. The short-term effects of a sexy tan are paid for by the long-term effects of a not-very-sexy wrinkled skin.

Even worse, many people risk immediate health damage while using tanning equipment. A survey by the British consumer's magazine *Which?* found that one in five users did not even wear the goggles which are essential to prevent eye damage.

WOOD SMOKE

The Issue
Bonfires and wood burning stoves are as traditional as sweeping up leaves in the autumn, Guy Fawkes night and Thanksgiving Day. For people in many countries, especially in the Third World, wood is still a vital source of energy. Yet despite its folksy, outdoors image, wood smoke is intensely dangerous to our health.

The Facts
Wood smoke is intensely carcinogenic and, breath for breath,

probably more dangerous than smoking cigarettes. It has been implicated in causing lung cancer. Wood smoke releases greenhouse gases and thus contributes to global warming. It also causes more immediate health problems such as smarting eyes and sort throats. Food smoked over wood also contains carcinogens. Although an occasional exposure probably won't do you much serious harm, it will certainly be uncomfortable if you get enveloped in smoke when the wind changes direction.

Assessment
All in all, wood fires are fairly unhealthy and environmentally polluting. To be avoided where possible, or kept for occasional festivals rather than a regular way of getting rid of garden rubbish. Bonfires are illegal in urban areas in some countries.

Minimizing Risk
For people using *wood stoves* as a regular source of heating, it is important to check that the flue is operating efficiently, and that as little smoke as possible escapes into the room. This means regularly sweeping chimneys, and ensuring that a flue is sealed completely to avoid leakage. Sealed stoves are probably better than open fires, and particular care should be taken if wood fuel is used in a room where people are sleeping. The wood will smoke less if it is properly dried, and if the fire is drawing well. Thus wood burners should collect the winter's supply of wood during the summer and ensure that it is well dried out before the wet and cold season. (This is also far more efficient from an energy and heating point of view.)

Most *bonfires* are made to burn twigs and leaves because, in traditional gardening, these are the bits of waste which take too long to compost. However, there are alternatives. Most municipal tips accept garden waste and so do some house to house refuse collectors, especially if they are neatly bagged or bundled.

Alternatively, if you have the space, leaves rot down to an excellent mulch in about three years. Three leaf mould bins, or just piles of leaves, can be used in rotation and ensure a constant supply of good mulch for use on the garden as a compost and a way of protecting plants against pest attack.

What Governments Could Do
There is a strong argument for both national and local authorities to introduce leaf mould composting as a safer option

than bonfires. These could be run on a municipal scale, accepting leaves from householders as well as those swept off streets and parks.

Further Information
Information about leafmould and other uses of leaves can be obtained from any good gardening organization, such as the Henry Doubleday Research Association and The Soil Association in Britain and the Rodale Institute in the United States.

PART FOUR: PESTICIDES

PESTICIDES

Pesticides are chemicals used for killing pests. There are literally hundreds of different pesticide chemicals, sold under several thousand trade names around the world. Pesticides are used in crop and livestock production, storing food, treating timber, keeping roads and paths clear of weeds, keeping canals and rivers navigable, protecting pets against fleas, getting rid of lice from hair, killing flies and thousands of other uses. Many of these are discussed in separate sections of this book.

There are a number of broad categories of pesticide, including:

- Acaricides: kill spiders and mites;
- Fungicides: kill fungi and moulds;
- Herbicides: kill weeds;
- Insecticides: kill insects;
- Nematicides: kill eelworms;
- Rodenticides: kill rats and mice.

There are several other general terms applied to pesticides, which are important to understand:

Chemosterilants: pesticides which stop insects from reproducing.

Formulation: the whole mixture containing the pesticide. It also includes some or all of the following: *solvents, adjuvants* (materials to make the pesticide stick to the plant); *dilutant* (usually water or oil); something to help the pesticide mix with the water.

Maximum Residue Limit: or MRL, the maximum permitted concentration of pesticide residue in or on a crop. International MRLs are set by the UN Food and Agriculture Organization and, to some extent, by the European Community.

Non-selective: a pesticide which kills a wide range of plants and animals.

Residual pesticide: a pesticide that can remain active in the soil, plants, wood, and so on for some time. (Also often called *persistent* pesticides.)

Selective pesticide: chemical which kills only one, or a small range, of plants and animals. The opposite is a general pesticide, which kills a wide range of plants or animals.

Systemic pesticide: an insecticide or fungicide taken into the plant so that pests attempting to feed on the plant, or moulds growing on the plant, are poisoned.

Translocated herbicide: herbicides which are taken into the plant at one point and carried (translocated) to other parts, where they cause damage, eventually killing the whole plant.

In this book, pesticides are discussed in a number of sections, including *garden pesticides* (see page 183), *pets* (see page 97), *pesticides in food* (see below), *spray drift* (see page 187) and *timber treatment* (see page 245). Sources of further information are given at the end of each of those. A general book which gives background information is *This Poisoned Earth* by Nigel Dudley (Piatkus Press).

In addition, there are some national and international organizations which can help with advice and information on pesticides. These are listed under *pesticides in food* (see page 182–3).

PESTICIDES IN FOOD

The Issue
In March 1988, the British government admitted: 'There is no such thing as pesticide free food.' Pesticides are used so widely that residues are likely to turn up, albeit in smaller quantities, even on food which has not been treated. The debate is now about whether these residues have any effect or not.

The Facts
Pesticides get into our food in a number of different ways:

Through the legal use of pesticides while crops are growing
Approval conditions tacitly assume that some pesticide is left on the food when it is sold and eaten. Providing the proper instructions are followed by the sprayer, the amount remaining should be within the official safety limits, i.e. the amount of pesticide which experts believe can be eaten without any ill-effects on health. In Britain this is the Acceptable Daily Intake (ADI). The

maximum amount of pesticide which falls within the ADI is defined by law in the European Community as the Maximum Residue Limits (MRLs). Britain finally fixed MRLs on some, but not all, of the pesticides used in 1989, although some other countries have had them in place for considerably longer.

Through illegal use of banned pesticides
Banned pesticides are still sometimes used and turn up on food. For example, *DDT* was banned in Britain in October 1984, but is still known to be available in places, and turns up as high residue levels on occasion. Illegal use of *aldicarb* on watermelon made over a thousand people unwell in the western United States in 1985.

Through illegal use of pesticides too close to harvest
The more toxic chemicals have a defined 'harvest interval', which is the gap which should be left between using the pesticides and picking the crop. If a grower breaks this harvest interval (often to ensure that the produce looks good in the shop) illegally high residues can still remain when the crop is eaten.

Post-harvest use of pesticides
Produce which is stored for long periods, such as grain, is often treated after harvesting. Recent research by the Ministry of Agriculture, Fisheries and Food in Britain found that over half the organophosphorus pesticide used on stored grain can survive the baking process and turn up in brown bread.

Accidental contamination by pesticides
Careless use of pesticide can result in *spray drift* (see page 187). If this takes place next to another crop, this can in turn be contaminated. Spray drift has affected a number of organic farms in Britain, meaning that the producers have temporarily lost their organic symbol of quality because of residues remaining on *organic food* (see page 59). It is also common when market gardens border directly onto arable farms growing cereals which are heavily sprayed.

Imports
Many imported foods are treated with chemicals which have been banned in the industrial countries. It is not uncommon for firms manufacturing chemicals which are banned at home to continue

to export them to countries with less stringent regulations. In 1980 it was estimated that 25 per cent of the US pesticide exports had been banned within America. Some of these hazardous pesticides come straight back to us in imported food – the so-called 'circle of poison'.

In Britain, pesticide residues have been monitored for some time, albeit not very thoroughly. A specialized sub-committee of the Ministry of Agriculture produces a report on pesticide residues every three years. In 1989, the report showed that pesticides had entered the food chain so thoroughly that virtually every food was likely to be contaminated: grains, vegetables, fruit, health foods, organic foods, dairy products, meat and even mother's breast milk. This was the cause of the quotation reproduced at the beginning of this section. It was the final culmination of a series of reports, from the Association of Public Analysts, the London Food Commission and the Home Grown Cereals Authority which all suggested that pesticides were extremely widespread in our daily diet. However, the Working Party on Pesticide Residues stressed that the amounts of pesticide detected were, on the whole, so minute as to be irrelevant. The argument had shifted subtly. No longer were we debating whether pesticides remained on food or not, but about whether it mattered.

The official attitude in Britain is still that it doesn't matter. However, in the United States there was a revolution in thinking about pesticide residues during the latter part of the 1980s, and this new perspective is starting to change attitudes all over the world.

There have been three key reports published in the United States which have brought about this shift in emphasis. The Environmental Protection Agency (an official body) has produced damning evidence suggesting that a number of pesticides, including daminozide, maneb, mancozeb and zineb, could cause cancer in people through residues left on food. For example, they calculated that the last three could cause an extra 123,000 cancers in the United States.

The National Academy of Science formed a committee to look into the pesticide residue issue and, in particular, into an apparent legal paradox created by the so-called Delaney Clause whereby the EPA is not supposed to set a tolerance limit (i.e. an ADI) for any pesticide which has been found to induce cancer in animals, and which concentrates in processed food. If taken literally, this would result in the banning, or at least suspension, of many older pesticides in America. The committee looked at risks from 28 pesticides identified as being particularly dangerous. They calculated that up to 1.46 million cancers could result, in time, from residues of these but that just ten pesticides accounted for 80–90 per cent of the danger, and that residues of these ten were concentrated on just fifteen foods. The foods are: grapes, tomatoes, potatoes, oranges, lettuce, apples, peaches, carrots, beans, soyabeans, corn, wheat, beef, chicken and pork.

Next, in February 1989, the Natural Resources Defense Council (a non-governmental organization) published a report which looked specifically at pesticide residue risks to children. The report concluded that between 5,500 and 6,200 of the current US population of pre-school children may develop cancer solely as a result of their exposure before the age of six to pesticide residues from just eight pesticides found on fruit and vegetables.

Again, it must be stressed that these figures are very preliminary. We still have an enormous amount to learn about risk assessment in these fields, and experts are arguing about orders of magnitude rather than fine tuning their understanding of risks. There are many specialists who still believe that pesticide residues at normal levels pose no significant problem. However, there are a growing number who do take them more seriously.

These new findings have caused a storm of protest throughout the industrialized world. One result in Britain was the public campaigning by a number of well-known people from the entertainment world and the formation of Parents for Safe Food by comedienne Pamela Stephenson. The initial focus of protests was the pesticide daminozide, sold under the trade name *Alar* and used on apples. Since then, concern has developed more generally about the use of pesticides (and a whole range of other 'food adulterants' in our diet.

Assessment
Even if the most pessimistic figures are right, there is still probably less risk from eating pesticide residues than, say, sharing a house with a regular tobacco smoker or drinking even a moderate amount of alcohol. But there seems to be increasing evidence that there are real, and measurable, risks which people may want to try and avoid.

Minimizing Risk
For people worried about pesticide residues, there is unfortunately often a trade-off between reducing traces of pesti-

cides and maintaining all the nutritious properties of the food.

- *Washing fruit and vegetables* will help reduce pesticides left on the surface. However, if 'systemic' pesticides have been used they will be incorporated into the main body of the plant and cannot be washed away;
- *Peeling fruit and vegetables* can also help reduce the concentration as many pesticides are concentrated on the skin. But so are many of the nutrients, which is why health experts have been telling us for years to leave the skin on. You have to make your own judgement about whether the risks are important enough to sacrifice the roughage and nutrients;
- *Cooking* also removes some of the pesticides, but again this also removes valuable goodness from the food;
- *Remove fat from meat* because some persistent pesticides are concentrated in the fatty tissue;
- Because of general environmental contamination, no food can be guaranteed 'pesticide free', but certified *organic food* is likely to have fewer residues than food produced by chemical farming methods.

Further Information

A number of books have been published on this issue. In Britain the Soil Association and Parents for Safe Food have produced a report which gives the background to the pesticide residue issues, called *Pesticides Under Our Skin*, it is available from the Soil Association, 86 Colston Street, Bristol BS1 5BB. The Natural Resources Defense Council's report *Pesticides in Our Children's Food*, by Mott and Broad was published in Washington DC in 1989. Andrew Watterson is writing a book which looks at common foods and what pesticides may be found on them, which is to be published by Green Print in Britain. And Parents for Safe Food have published *The Safe Food Handbook* (Ebury Press, 1990) which gives a useful guide to this and other issues. There are also many more general books on pesticides which stray into this area or give valuable background, some of which are listed in the *spray drift* section (see page 187) or in the bibliography at the back (see page 265).

There are a lot of research and campaigning groups involved in pesticides. Some are listed in the resources section at the back of the book. The Pesticide Action Network (PAN) has offices in Europe, North and South America, Asia and Africa. In Britain,

The Pesticides Trust (20 Compton Terrace, London N1 2UN) has information available. And Parents for Safe Food, which was started in Britain, now also has offices in Australia and New Zealand. In the United States, the Natural Resources Defense Council has carried out the bulk of work on pesticide residues, but there are many other citizens groups involved in this issue. And The Soil Association in both Britain and Australia will be able to give advice about alternative supplies of food.

PESTICIDES IN THE GARDEN

The Issue
Over a hundred different types of garden chemical are available over the counter, are sold to people of any age, and are promoted by strong advertising campaigns. These pose risks to health if misused, or if allowed to drift over areas where people are living, cooking food or children are playing. They can also contaminate the user.

The Facts
All domestic pesticide applications have their own hazards. These risks are further increased because most amateur users know little about the potential dangers involved.

There are an enormous number of garden pesticides available. Every year, Britain's amateur gardeners spend about £15 million on pesticides to use in their gardens. There are well over 500 products on the market, made up from more than a hundred different insecticides, molluscicides, fungicides, herbicides and rat poisons, mixed in a variety of formulations. The Soil Association calculates that about a kilo of active pesticide ingredient is used, on average, on every acre of British back garden.

Spraying pesticides in back gardens poses a number of specialized health and safety problems, including:

- Use of chemicals by untrained people;
- Use of chemicals close to houses, where clothes are drying, on vegetable gardens and in places where children are likely to be playing;
- Storage of chemicals where children have easy access;
- Damage to wildlife which may be specifically attracted to the garden by bird tables, nest boxes and so on;
- Risk to pets.

Government and industry representatives always stress that, if used correctly, garden chemicals don't pose any threat to health. However, although some of the most hazardous chemicals are, indeed, not released for general sale, a survey by the Soil Association found a substantial number of hazardous chemicals for sale over the counter in chemists and garden centres in Britain. (Regulations vary between countries about what is available for home garden use, but these figures probably apply broadly for other countries as well.) The Soil Association found the following for sale in garden shops and centres:

- 38 chemicals irritating to the eyes, skin or respiratory tract, including *benomyl, carbendazim*, and *glyphosphate*;
- 25 chemicals which are known or suspected carcinogens. These included *captan, trichlorfon* and *dicoful*;
- 29 chemicals which are known or suspected mutagens or teratogens, including *malathion* and *MCPA*;
- A number of chemicals which are acutely poisonous, such as *paraquat* and *nicotine*.

A table of garden pesticide hazards, produced by the Soil Association, follows on page 182. (Note that not *all* the pesticides listed will be available in every country.) Each chemical is likely to be sold under a range of different trade names. This means that there are literally hundreds of hazardous pesticide products being sold for use by householders in their own gardens. To use the table, look for the chemical name, or names, which should also be displayed on the packet.

Garden pesticides are usually supplied in very small amounts as compared to those used on farms or for municipal pest control. However, this advantage is at least partially offset by the lack of training for people using garden pesticides, and the fact that the pesticides are used next to where people live.

There is tacit acceptance that very hazardous products have been available in the recent past. In 1987, weedkillers containing *ioxynil* and *bromoxynil* were banned from home garden use in Britain after research suggested links with birth defects. However, products containing these chemicals can still be found in many garden shops, despite the ban.

Although all advertisements are obliged to refer users to the safety instructions on the bottle or packet, the survey found a number which actually showed photographs or drawings of people using pesticides in a dangerous way. For example, one photograph showed a woman holding a spray nozzle at the same height as her face and spraying tree foliage, thus risking spray blowing into her face.

In the past, there have been some tragic accidents where pesticides stored in old drink bottles have been drunk by children, who have mistaken them for soft drinks. (Some look quite like cola for example.) Today, there are legal requirements to store pesticides in their correct container in Britain, complete with warning label. However, there is no obligation to keep pesticide in childproof containers, in the way that most medicines are now controlled. Nor do most pesticide advertisements or labels stress that pesticides should be kept in locked cupboards and out of reach of young children.

Assessment
People undoubtedly do get affected by pesticides in the back garden. The commonest effect is probably something like a summer cold, or hay fever, i.e. a sore throat, streaming eyes and itchy skin. Many people would not connect this with pesticide use at all. The risks of longer-term effects are probably very small for normal use, but are certainly not zero, so that it is worth taking care with pesticides or, ideally, dispensing with them altogether.

Minimizing Risk
Back garden pesticide users could reduce the risk to their own health, and the health of their families, by a number of simple steps. These include:

- Always wearing gloves when handling or using pesticides, and keeping one pair of gloves solely for pesticide use;
- Wearing a mask when spraying pesticide, and never spraying in windy conditions or when other people are nearby;
- Shutting windows, and removing washing, deckchairs, children's toys and, of course, food and drink from the garden before starting spraying;
- Washing hands thoroughly after spraying or handling pesticides;
- Always storing pesticides in the correct containers, and in a locked place where children cannot gain access;

Continued on page 186

	Alloxydim sodium	Amintriazole (Amitrole)	Ammonium sulphamate	Asulam	Atrazine	Benomyl	Bioresmethrin	Bromophos	Bupirimate	Captan	Carbaryl	Carbendazim	Chlordane
CLASSIFIED AS POISON													
BANNED OR RESTRICTED IN SOME COUNTRIES		•				•					•		•
OTHER HEALTH EFFECTS		•			•	•	•		•	•		•	•
RESISTANCE DEVELOPED IN SOME PESTS	•					•	•		•	•	•	•	
PERSISTENCE LONG = L MEDIUM = M					M	L	M		M	L	M	M	L
SPRAY DRIFT PROBLEM AND/OR HARMFUL TO PLANTS		•	•	•	•					•			
DANGEROUS TO BEES AND OTHER INSECTS						•		•	•	•	•		•
DANGEROUS TO FISH						•			•	•	•	•	•
DANGEROUS TO ANIMALS AND BIRDS						•		•					•
HARVEST INTERVAL (in days)	28 56							7	1–14		7–36		
IRRITATING TO EYES, SKIN AND/OR RESPIRATION	•	•	•		•	•			•	•	•	•	•
ANTI-CHOLINESTRASE COMPOUND											•		
SUSPECTED OF CAUSING BIRTH OR GENETIC DEFECTS		•			•	•				•	•	•	•
SUSPECTED OF CAUSING CANCER		•				•				•	•	•	•
CLASSIFICATION	H	H	H	H	H	F	I	I	F	F	FI	F	I

A = ACARICIDE F = FUNGICIDE H = HERBICIDE I = INSECTICIDE
M = MOLLUSCICIDE

Chloroxuron	Copper	2,4-1	Dalapon	Derris	Diazinon	Dicamba	Dichlobenil	Dichlofluanid	Dichlorphen	Dichlorprop	Dichlorvos	Dicofol	Dimethoate	Dinocap	Diquat	DNOC	Fenitrothion	Fenarimol	Fenoprop	Ferrous sulphate	Folpet	Formothion	Gamma HCH
											•					•	•						
		•												•									•
								•		•	•	•	•	•	•	•	•					•	•
					•							•	•	•									
	L			M	M					M	L	M	M	L	M	M					L	M	M L
	•	•	•		•	•	•			•	•		•					•	•		•		
					•	•					•	•	•	•		•	•					•	•
	•	•		•	•	•	•	•	•	•	•	•	•	•		•	•	•				•	•
	•				•						•				•	•	•						•
			1		14			3–21			1		2–21	7	7	7–14							14
•	•	•			•	•	•		•		•	•	•	•	•	•	•	•	•		•		
					•						•			•		•							
	•				•	•			•		•	•	•			•				•	•	•	•
	•										•	•											•
H	F	H	H	I	I	H	H	F	H	H	I	A	I	F	H	I	I	F	H	H	F	I	I

From '*How Does Your Garden Grow?*' by Nigel Dudley (The Soil Association 1986).

	Glyphosphate	Heptenephos	Ioxynil	Malathion	Maleic hydrazide	Mancozeb	Maneb	MCPA	Mecoprop	Mercurous chloride	Metaldehyde	Methiocarb	Nicotine
CLASSIFIED AS POISON													•
BANNED OR RESTRICTED IN SOME COUNTRIES							•						•
OTHER HEALTH EFFECTS		•		•		•				•	•	•	•
RESISTANCE DEVELOPED IN SOME PESTS				•									
PERSISTENCE LONG = L MEDIUM = M	L			M		L	L			L	M		
SPRAY DRIFT PROBLEM AND/OR HARMFUL TO PLANTS	•		•	•				•	•				
DANGEROUS TO BEES AND OTHER INSECTS		•		•			•			•	•	•	•
DANGEROUS TO FISH	•	•	•	•	•		•			•	•	•	•
DANGEROUS TO ANIMALS AND BIRDS	•	•								•	•	•	•
HARVEST INTERVAL (in days)	7	1		1		7	7–14				10	7	2
IRRITATING TO EYES, SKIN AND/OR RESPIRATION	•		•	•		•	•	•	•				
ANTI-CHOLINESTRASE COMPOUND		•		•								•	
SUSPECTED OF CAUSING BIRTH OR GENETIC DEFECTS			•	•	•	•	•	•					•
SUSPECTED OF CAUSING CANCER					•		•	•					•
CLASSIFICATION	H	I	H	I	H	F	F	H	H	F	M	M	I

A = ACARICIDE F = FUNGICIDE H = HERBICIDE I = INSECTICIDE
M = MOLLUSCICIDE

Oxycarboxin	Oxydemeton methyl	Paraquat	Permethrin	Picloram	Pirimicarb	Pirimiphos methyl	Propachlor	Propincanazole	Pyrethrum	Quintozene	Resmethrin	Simazine	Sodium chlorate	Sodium chloracetate	Sulphur	2, 4, 5–T	Tar oil	Tecnazene	Thiophanate methyl	Thiram	Trichlorfon	Zineb
	•	•															•					
		•														•			•			
		•	•		•	•		•			•	•		•		•	•	•				•
•									•		•							•	•			
M				L		M				M	L	M	L	L	L	M			L	M	M	M
		•	•	•			•			•	•	•	•	•	•	•			•			
	•		•		•	•				•	•		•									•
	•		•	•		•	•	•			•				•	•	•	•	•	•	•	
	•	•			•								•		•							
	14–21				2–21	7–21		28										14–36		7–21	2	2–21
		•	•	•		•	•	•	•	•		•	•	•	•	•	•	•	•	•	•	•
	•				•	•															•	
		•			•	•				•	•				•					•	•	•
			•	•											•	•	•	•	•	•	•	•
F	I	H	I	H	I	I	H	F	I	F	F	H	H	H	F	H	I	F	F	F	I	F

From '*How Does Your Garden Grow?*' by Nigel Dudley (The Soil Association 1986).

- Wearing long trousers, and a long-sleeved shirt or jacket, when spraying pesticide, and avoiding spray drifting into your face;
- Spraying fruit and vegetables in the early morning or late evening, when there are fewer flying insects about;
- Putting slug pellets in places where birds and hedgehogs cannot reach them, and clearing away any dead slugs you see (which will still contain the poison);
- Always reading the label on pesticide containers before use.

It is also worth noting that pesticides are not necessarily very effective in any case. A survey published in 1988 in *Gardening Which?*, a British consumer's magazine, found that only two of the products out of a wide range tested kept a garden path free of weeds as advertised. The rest were, at best, temporarily effective and some did not clear all weed species.

A much more satisfactory answer is to cut pesticides out altogether. Thousands of gardeners already practise *organic growing methods*, where pests are controlled by biological and cultural techniques rather than chemicals. These include encouragement of predators of pests, such as ladybirds to control aphids; using barriers on fruit trees to stop insects from climbing up the trunk; practising crop rotation to stop crop-specific pests from building up in the soil; and companion planting of different species to confuse pests which hunt by scent. Organic gardeners also use a few, plant-based, insecticides to control occasional pest outbreaks in an emergency.

Further Information
Information on organic gardening is available from the Henry Doubleday Research Association (Ryton on Dunsmore, Coventry) and the Soil Association (86 Colston Street, Bristol, BS1 5BB).

US organic groups include Natural Food Associates (PO Box 210, Atlanta, Texas 210, USA) and the Rodale Institute which publishes a very wide range of books. In Australia, get in touch with the Soil Association of South Australia Inc (GPO Box 2497, Adelaide, South Australia 5001) for local contacts. For a list of organic groups throughout the world, write to IFOAM (The International Federation of Organic Agriculture Movements (c/o Ookozentrum Imsback, D 6695, Tholey-Theley, Germany).

SPRAY DRIFT

The Issue
Everyone is worrying about pesticides today. But are they really a problem? Is it dangerous to go walking in the country? And what can you do if a sprayer allows pesticide to drift across your land?

The Facts
'Pesticide' is a general term used to describe a whole range of chemicals including herbicides, fungicides, insecticides, acaricides (which kill spiders and mites) and so on. There is considerable debate about how dangerous pesticides are to humans, which you can find summarized in the section on *pesticide residues on food* (see page 174). One of the most dangerous aspects of pesticide use is the risks which occur when pesticides are applied. The best known of these is the risk of pesticide drifting away from the target area and contaminating surrounding wild areas, gardens, farms, roads or public places. This is known by the general term of spray drift, and includes three main effects:

- *Spray drift* or *drop drift*, which occurs when small pesticide droplets are blown off target by the wind. This is the most important type of spray drift, and will be described in more detail below;
- *Vapour drift*, which is the drifting of vapour from pesticides which evaporate after being applied. This can occur two or three days after spraying, which makes it very difficult to trace where the problem came from. It used to be a fairly serious problem in some areas where particular volatile (i.e. easily evaporated) herbicides were used, although voluntary bans have helped cut down the problem to some extent;
- *Blow*, which refers to the blowing of granular pesticides away from where they are supposed to be by high winds. This is reckoned to be the least important of the three, although it has never been studied in much detail. Blow is likely to be especially hazardous if it does occur because pesticides applied as granules are often amongst the most dangerous.

All types of spray drift result in pesticide ending up in the wrong place specifically where the chemical can damage wildlife, pets and people. Examples of how this can happen include:

- Spray drifting over a garden, contaminating vegetable plots, grass where people sit and children play, or blowing directly into a house (see *garden pesticides* page 179);
- Spray drifting to the edges of fields and onto vegetation next to a public footpath. This can cause skin irritation to people walking along in shorts, skirts or with bare arms;
- Spray drifting over public roads and contaminating drivers, cyclists and walkers;
- Spray from aircraft (aerial spraying) which can often cover a wider area because of the greater speed with which the plane travels, and thus the smaller margin for error.

All these things really happen. Official figures used to make it sound as if it was a very minor problem, but groups like the Soil Association and Friends of the Earth have collected detailed case studies. The 'official' figures for complaints about spray drift were about 60 in 1985. But in one public meeting that I attended in Louth, Lincolnshire in the same year, at least thirty people complained of *personally* having suffered spray drift. The problem is far larger than is generally realized.

The health risks are very difficult to assess. Again, official figures play down any risks, while some unofficial organizations, and a growing number of doctors, believe that there are substantial risks. These include both immediate risks of poisoning – which can result in local irritation or more serious illness – and the risks of chronic illness from exposure to a carcinogen or teratogen. There is now a new organization, the Pesticide Exposure Group of Sufferers which is collecting data on these incidents.

One of the problems is that the level of reporting is very low, and the number of people who even recognize that an illness may be caused by pesticide exposure in even lower. One survey carried out by the Open University in Britain included interviewing 80 farmers. Forty-one of these admitted that they believed they had suffered ill-effects from exposure to pesticides, but only one had reported the incident to the authorities.

Often the effects of serious drift can be seen, either because the spray is spotted drifting across or by spotting the effects on foliage. A swathe of browned or dying vegetation leading up to an area that has been sprayed is a good indication of drift. But it must be remembered that many pesticides, including insecticides and some fungicides, will not leave any very clear signs that they have passed by.

Another spray drift issue, which has persistently been ignored by both scientists and the media to date, is the chronic drift of very small droplets. Measurements of spraying machinery at the Centre for Tropical Pest Control in Ascot, near London, found that up to 20 per cent of the pesticide was released as tiny droplets less than a hundred microns in diameter. These are liable to drift whatever the weather, and could stay in the atmosphere for a very time. If diluted with water, this will evaporate leaving an even smaller particle of pure chemical. Particles of the same size are known to drift up into the stratosphere. Many of these chemicals are highly toxic carcinogens, mutagens, or have effects on the immune system. As far as I can find out, after fairly extensive enquiries, no work at all has been carried out on the possible health impacts of constant exposure to very low doses of a whole cocktail of different chemicals existing as a background in the air.

Assessment
It is still very difficult to make an assessment, because the research hasn't been done. There is a real risk of irritation of various sorts to skin, lungs and eyes if you are caught in the spray drift of many pesticides. But because these symptoms may often be caused by pollen or other allergens, it is usually difficult to prove that pesticides were involved. A few unfortunate people suffer serious illness as a result. There is also sufficient evidence to justify concern about longer-term effects, but little hard scientific evidence one way or the other as yet.

Minimizing Risks
Beware. Getting involved in issues of health effects of pesticides means entering a quagmire of conflicting ideas. There is a lot of money and emotion tied up in chemical farming and everyone who claims that they have suffered ill-effects can expect a good bit of rather prejudiced ridicule. A few years ago two incidents occurred at which illustrate this point. In one, a group of girls marching through a Yorkshire town in a band all became ill at the same time, with coughing fits and faintness. Spots on their clothes led to suspicions of pesticide contamination, but there was no conclusive evidence and the police said it was a case of mass hysteria. Around the same time, in Hertfordshire, a group of part-time firemen were sprayed by a plane and all ended up in hospital. No one accused them of mass hysteria. I don't have any proof one way or the other about the first case, but it does illus-

trate a general point about the problems people face when they have to report a pesticide incident, or something similar.

That said, there are a number of steps which can and should be taken.

Avoiding Risk
It is difficult to avoid spray drift risk if you live or travel in areas which are being sprayed. But one precaution worth taking is to check whether a field has just been sprayed before walking through on a footpath and brushing against vegetation. The UK Ramblers Association are lobbying for it to be illegal to overspray public footpaths.

In addition, if you live next to a farm, you could try asking the farmer to let you know when he or she is going to spray a field, so that you can keep children and pets inside, and watch out for drift. Some people have had success here, but there is no legal obligation on the part of the farmer.

If a Spray Drift Incident Occurs
There are a number of things which you should do immediately if drift occurs over you or food in your garden:

- Follow the drift upwind to find its source;
- Inform the sprayer that they (or their employer) will be held responsible for any damages occurring as a result;
- Photograph and record any obvious signs of damage to the surroundings – vegetation and so on – as evidence;
- If you think you have been contaminated, get a blood test within 24 hours (or 48 hours if this is the earliest possible). Some places which carry out blood tests are: The Lister Hospital, Allergy and Environmental Medicine Department, Chelsea Bridge Road, London SW1W 8RH (071-730-3417). The Breakspear Hospital, Abbots Langley, Hertfordshire, WD4 9HT (09277 61333).
- Find out what sort of pesticide was used and check in reputable reference sources to find out any details about health effects. Some good books are listed at the end of this section;
- Also use reference sources to check the harvest interval for the pesticide, which is the gap which must be left between spraying crops and eating them, if any vegetables or fruit have been affected.

This isn't by any means total protection, but it should help. If you think that you have suffered health effects, consult your doctor or an independent specialist. It is possible but – in many countries including Britain – extremely difficult to claim successfully against a pesticide user for health effects which you may have suffered.

Further Information

One of the best books available for information about pesticide hazards is the *Pesticide Users Health and Safety Handbook* by Andrew Watterson, which is published by Gower. A general guide is *The Pesticide Manual* by Peter Hurst, Alastair Hay and Nigel Dudley (Pluto Press). The Soil Association (86 Colston Street, Bristol, BS1 5BB) also has a great deal of information available on spray drift and its problems, including two pamphlets called *Pall of Poison* about spray drift, and *Drifting into Trouble* about claiming against pesticide damage.

PART FIVE: TRANSPORT

CAR ACCESSORIES

The Issue

Everyone knows that vehicle pollution is a major health and environmental problem (see page 202). But a number of the accessories, replacement parts, sprays and liquids used by car drivers are potentially dangerous as well. This section gives a quick overview of some things to watch out for and suggests general ways in which the motorist can reduce risks of contamination.

The Facts

Most of the potential hazards involving cars are fairly clear cut; it's obvious that a lot of the materials you're dealing with shouldn't be eaten or drunk! But there are a few more hidden dangers, from vapours, dusts and from contamination through the skin. There are also some things which are more toxic than others, and need treating with extra special care.

Anti-freeze

Many anti-freeze liquids contain *ethylene glycol* which is toxic and a skin irritant. It can have effects on the nervous system and potentially fatal liver damage. Avoid getting it on your hands and don't inhale any vapour. It is flammable.

Battery Acid

Batteries contain *sulphuric acid* in a fairly concentrated form. This is one of the most poisonous and corrosive acids and you should always wear gloves and old clothes when handling it. If any does splash onto skin or – especially – eyes, wash it off immediately with plenty of cold water. If eyes are contaminated you should consult a doctor at once. And beware of the white, powdery material which builds up around the terminals, because this is toxic and corrosive as well.

Brake and Clutch Fluid

This often contains *polyalkylene glycol ethers* which are slight skin irritants and should not be swallowed. Mineral oil-based fluids are also available but, again, these can cause skin irritation.

Brake and Clutch Linings

Linings have traditionally been made from *asbestos*, which is extremely toxic (see page 85). If you have to change or handle these yourself, wear gloves, and carry out the operation in the open air, taking care not to breath in any of the dust. Old brake linings should be wrapped in newspaper or a bag and put in the dustbin. Non-asbestos linings are now available and should be chosen wherever posible.

Damp Starts
Many damp starts are available, including some in *aerosols* (see page 83). These all contain *solvents* (see page 241) and some are hazardous, so you should avoid breathing in spray or vapour. Solvents used include *xylene* and *trichloroethylene*.

Engine Cleaners
Many engine cleaners contain *cresylic acid* which is highly poisonous. You should also avoid skin contact, because there is a risk of contamination through the skin. It is highly corrosive, can have effects on the central nervous system and kidneys and has a harmful vapour.

Gasoline
See *petrol*

Grease
Despite being such a traditional material, grease is more hazardous than many people realize. Problems are particularly common with old grease, which is likely to be contaminated with fine particles of metal – in this state some people find it a serious skin irritant.

Lubricating Oil
Oil is also a skin irritant. Old oil is likely to contain lead and polycyclic aromatic compounds, both of which are carcinogenic.

Petrol (Gasoline)
Petroleum is a toxic liquid and carcinogen. This may be a problem for people working in garages, and higher than average lung cancer levels have been measured in people who work as petrol pump attendants for along time. Petrol is discussed in more detail under *vehicle pollution* (see page 202).

Underbody Treatments
These are used to seal the underside of the car against rust and so

on. Most are based on either *wax* or *bitumen*. Both these are likely to contain *solvents* and bitumen is also itself carcinogenic. You should avoid skin contact, and should not inhale spray or vapour.

Windscreen De-icers
These often contain *methanol*, which is a cumulative poison which can be ingested through the skin or by breathing, and is also flammable. It can affect eyes, causing blindness in extreme cases. It is a cumulative poison which is especially serious for people exposed on a daily basis.

Windscreen Wash
Special compounds are sold to mix with water in windscreen washes, both to avoid freezing and also to improve cleaning ability. These often include *ethylene glycol* or *ethanol* and *ammonia*. These are all poisonous or irritant.

Assessment
Not surprisingly, many of the materials used in cars are fairly unpleasant. However, some standard precautions, as outlined below, should be able to reduce the risks very considerably.

Minimizing Risk
The main risks of contamination come while working on the car in a garage, and the people most likely to be affected by chronic pollution are professionals who are in constant contact with toxic materials. However, there can be some short-term irritant effects and, as explained in the introductory section, it is worth trying to minimize any contact with carcinogens or teratogens.

- Always wear gloves when handling car accessories;
- Carry out all but very minor jobs either outside or in a well-ventilated garage;
- Avoid breathing any sprays or vapours as far as possible. It is better to choose non-aerosol options wherever possible;
- Never smoke when working on a car, and avoid naked flames, as many of these liquids and vapours are highly flammable;
- Take extreme care with brake and clutch linings containing asbestos.

Further Information
C for Chemicals by Michael Birkin and Brian Price (Green Print)

gives additional details about many of these substances.

CYCLING AND JOGGING

The Issue
More and more people are taking up cycling and jogging for health reasons. But do they do themselves more harm than good by breathing in polluted air?

The Facts
There are lots of reasons for being cautious about hard exercise like running and cycling from a general health point of view – these can include anything from sprained muscles to heart attacks, and there are large question-marks about the long-term implications for backs and knees of running on hard surfaces. These are beyond the remit of this book, but anyone thinking of taking up running should get proper advice about equipment and, if they're over thirty, probably get medical advice.

What we're concerned with here is whether exercise which makes you breathe deeply and rapidly substantially increases risks of ill-health from breathing in the pollutants discussed under *air pollution* (see page 151). The short answer is that no one really knows. There have been several studies in very polluted areas, such as Los Angeles and inner London, with conflicting results. Some have found that carbon dioxide in the lungs of runners (used as a measure of how much pollution has been breathed in) are no worse than in other people; others end up recommending people never to run in the city!

We do know that there are good reasons for being wary about the health effects from *vehicle exhausts* and it seems reasonable to assume that someone taking exercise which increases breathing rate would take in correspondingly greater amounts of pollution. But whether this is a significantly greater proportion overall, and whether it is more than compensated by the health-giving effects of exercise, is still an open question.

Minimizing Risk
If you are going to run or cycle, there are some fairly obvious precautions which can be taken to minimize any risks:

• Don't run along busy roads, especially those where cars are likely to be idling and thus releasing more exhaust fumes;

- Don't run in fogs or mists which trap pollution;
- Try to run in the early morning or late evening when there is less traffic about;
- Cyclists could try wearing a face mask, although this will only keep out some of the larger particles, like dust. A better bet for avoiding pollution is probably to find routes which avoid the busiest roads.

DIESEL ENGINES

The Issue

Diesel engines are used in virtually all lorries and buses in Europe and in an increasing number of private cars. Diesels are major sources of black smoke and a range of pollutants, and are probably more dangerous to human health than petrol (gasoline) engines. European legislation is lagging far behind that in the United States, despite there being far fewer diesel engines in America.

The Facts

Virtually all new lorries and buses in Europe now run on diesel engines. In addition, about 15 per cent of new cars are also diesel-powered, although this figure varies greatly between countries. In Britain only about 4 per cent of new cars run on diesel, while in Italy the figure is nearer 26 per cent. Some analysts believe that almost a quarter of cars produced in Europe by the end of the 1990s will be diesel-powered. The traditional 'black cabs' driven in British cities are diesel-powered.

Diesel engines produce a different mix of pollutants to petrol-driven cars. They have a number of clear advantages. Diesel engines do not emit *lead*, and generally have a low pollutant emission – if they are properly maintained and tuned. Unfortunately, this often isn't the case, whereupon emissions rise steeply and they become highly polluting.

Diesel engines are a major source of black smoke and can be responsible for up to 90 per cent of black smoke in urban areas. They also produce more *polycyclic aromatic hydrocarbons* (PAHs) and *sulphur dioxide* than petrol cars, about the same amount of *nitrogen oxides* and less *carbon monoxide*.

Over the last few years, a number of studies have looked in

detail at the health effects of diesel engines, although there have been few epidemiological studies looking specifically at the impact of diesel fumes on the population. The following general conclusions have been drawn.

Smoke
Diesel smoke aggravates diseases such as bronchitis, asthma and cardiovascular problems.

Cancer
Diesel engines can emit polycyclic aromatic hydrocarbons (PAHs) which are strongly suspected of causing cancer. This has been backed by animal studies using diesel emissions. One study using mice found that the mixture of PAHs from a diesel exhausts was highly active in initiating tumours. Studies on animals also suggest that diesel could add to the risk of cancers developing in those exposed to other carcinogens, such as cigarette smoke.

This has been borne out by studies of workers and others who have been exposed to high levels of diesel fumes, although in most cases sample numbers are too low to be statistically significant (see page 13 for discussion about how risk assessments are made). There is still no conclusive proof, but there is good evidence that there *could* be a link between diesel pollution from vehicles and development of lung cancer.

Mutagenicity
The National Swedish Environment Protection Board has concluded that the mutagenic effect of diesel engines is about an order of magnitude higher than for petrol engines. These conclusions are broadly similar to those obtained by the US Environmental Protection Agency and in Japan. The Dutch Research Institute on Road Vehicles has tentatively suggested that the mutagenic activity from diesel exhausts could be six times higher than from petrol exhausts.

The US government has concentrated control measures on buses rather than lorries, because they argue that the public are more exposed to heavy fumes from buses in towns and cites.

These findings don't mean much on their own, unless we also have some idea of the overall importance of *air pollution* to human health. This has been the subject for much disagreement, and is discussed in more detail in the section on air pollution (see page 151).

At the moment, although European countries have far more diesel engines on the road than the United States, European regulations regarding emission control are nowhere near as strong. Some would argue that this is the *reason* for lax regulations. Despite many opinion polls showing that the public does not like diesel smoke, and clear evidence about health risks, the EC regulations regarding particulate emissions from diesel engines are about three times weaker than those in the United States.

Assessment
There seems to be good reason for concern about the emissions from diesel engines, and for trying to avoid them as much as possible. This is especially true as the means for reducing emisions do exist and are already being employed in some places.

Minimizing Risk
There is not very much the individual can do, except get out of town... However, young children, the sick and the elderly are thought to be especially susceptible to some of the side effects of diesel fumes, so it might be worth trying to avoid the places where, for example, buses and taxis are idling in a narrow street, and thus liable to build up large concentrations of diesel smoke. If you can see a lot of smoke coming from a diesel engine than the chances are that quite a lot of other toxic pollution is being emitted at the same time.

What Governments Could Do
The technology is available to meet the more stringent emission controls with only fairly minor modifications to the diesel engine. There is no excuse for governments in Europe not to follow the lead set by the Americans and tighten up controls on their own diesel vehicles.

Further Information
Much of the material for this section was drawn from two overview documents prepared for Friends of the Earth in Britain; *Air Pollution from Diesel Engines* and *Particulate Pollution from Diesel Engines,* both by Claire Holman. See also *The Toxicity of Diesel Emissions* by R. E. Waller (Warren Spring Laboratory, 1986, UK) and *Health Risks Resulting from Exposure to Diesel Exhausts: Impacts of Diesel-powered Light-duty Vehicles,* National Research Council (Washington 1981). Claire Holman's reports

contain full reference sections. See also a report from the National Society of Clean Air, *Dirty Diesels*, for a succinct overview of the problem (NSCA, 136 North Street, Brighton, Sussex).

There are an increasing number of publications dealing with environmental impacts of road transport; some of the most useful are listed in the section on *vehicle pollution* (see below).

VEHICLE POLLUTION

The Issue
Vehicle pollution contributes to urban smogs, acid rain and forest dieback. Some of the pollutants also have significant effects on human health.

The Facts
Motor vehicles cause pollution because the internal combustion engines which power them can never be 100 per cent efficient. As a result, some of the products of incomplete combustion are emitted in the exhaust. In this section, we look at engines powered by petrol (gasoline); *diesel engines* are examined in a section of their own (see page 199). Pollutants from petrol-driven road vehicles posing a known risk to human health include carbon monoxide, hydrocarbons and nitrogen oxide.

Carbon monoxide
Carbon monoxide is an extremely poisonous gas, and is usually what kills people when they commit suicide by running their car engine in a confined space. It also has a number of important, sub-lethal, effects on health. Carbon monoxide acts by depriving the body of oxygen, which at lower concentrations causes impaired perception and thinking, headaches, slowing down of reflexes and drowsiness. Carbon monoxide at sub-lethal levels is a major air pollutant in some towns and cities. Road transport is by far the largest single source of carbon monoxide in most countries. In Britain, for example, over four and a half million tonnes of carbon monoxide are emitted from road vehicles (1988 figures), making up 85 per cent of total emissions.

Hydrocarbons
Hydrocarbons come from unburnt or partially burnt fuel. They

have a number of direct effects on health, and are also one of the constituents of *ozone*, which is itself damaging to human health, causing coughs, impaired lung function, irritation of the eyes and breathing passages, and a tendency towards headaches (see also *air pollution* (see page 151). Road transport contributes 30 per cent of the total British emissions of hydrocarbons, emitting something over half a million tonnes a year.

Nitrogen Oxides

Nitrogen oxides include both *nitric oxide* and *nitrogen dioxide*. The latter causes a range of health risks, including increased susceptibility to viral infections, lung irritation, and increasing the chances of susceptible people developing bronchitis and pneumonia. Asthmatics may be especially sensitive to nitrogen dioxide. Over a million tonnes of nitrogen oxides were emitted by British cars and lorries in 1988, making up almost half the UK total of these pollutants.

Assessment

There is still a great deal of debate about the precise risks involved from motor vehicles and most international studies remain cagey about overall effects (although see the section on *air pollution* see page 151).

Further Information

There are a growing number of reports and books on environmental effects of air pollution, although relatively few concentrate directly on health. *Transport and the Environment*, from the OECD gives a good overview (OECD 2 rue Andre-Pascal, 75775 Paris; available from HMSO bookshops in Britain).

PART SIX:
OTHER HAZARDS

BATTERIES

The Issue

Batteries are an indispensable part of life for many people in the 1990s, giving access to stereo headphones, automatic cameras, watches, clocks and a host of other electronic accessories. Some companies have started marketing 'green' or 'environment friendly' batteries. What is wrong with the others? Do they pose a risk to users?

The Facts

The current greening of battery manufacture is aimed more at protecting the general environment than the individual, both from the perspective of overall energy and resource use, and by avoiding some of the most toxic materials used in battery manufacture. As such it is certainly to be encouraged. However, in terms of immediate effects on human health, there is probably little to worry about unless the battery is burnt or punctured.

Standard batteries are made of zinc-carbon. The most toxic materials they contain are usually the battery acids, which are poisonous and corrosive. Some of the small 'button cell' batteries used in cameras and hearing aids contain mercury. Rechargeable batteries, which are more environment-friendly from an energy and resources point of view, actually contain slightly more dangerous materials in the form of nickel and cadmium. There are also a growing number of batteries marketed as specifically environment friendly which have avoided some of these materials, especially cadmium.

In normal use, none of these should come into contact with the user. However, there are a number of exceptions:

- If batteries are burnt they can release toxic fumes and, in a few cases, also explode. They should never be put in a fire or boiler. If there is no recycling point nearby, they are probably best put in the dustbin;
- Puncturing the case of a battery can also release toxic materials. The risk of this is likely to be greatest with larger batteries which are encased in metal and card unlike the slim metal batteries used in torches and electronic equipment;
- If a battery is left for a long period of time without use, the acids inside will gradually eat away at the case and release any toxic materials inside. If this happens, don't touch the battery,

or any of the crusty white stuff which will have collected outside, as both will be corrosive.

It is easy to cut down on batteries. Using rechargeable batteries
or a mains transformer is both kinder to the environment and
considerably cheaper in the long run.

What Governments Could Do
The greatest risk posed by batteries is through their disposal in
municipal dumps, *waste treatment plants* (see page 243) and other
places, where they add to the general pollutant load. Legislating
for proper recycling of batteries would allow almost all the most
dangerous substances to be used again and again, thus enormously reducing the amount of pollutants from battery use.

CARCINOGENS

The Issue
A carcinogen is something which increases the risk of developing
cancer. Testing for carcinogenicity is difficult, and even once
something is known or strongly suspected of causing cancer, as
explained in the introductory section, it is difficult to assess the
level of danger involved.

The Facts
There are many, many known or suspected carcinogens. A lot are
discussed in this bok. They can include things which occur naturally, such as *sunlight* (see page 166), *aflatoxins* (see page 29),
radon gas (see page 138), and *bracken* (see page 155), along with
a vast number of artificial chemicals or manufactured products,
including well-known hazards like *dioxins* (see page 211) and
asbestos (see page 85). The situation is further complicated
because a harmless substance can sometimes be changed into a
harmful, or potentially harmful, substance either through an
intermediary chemical (a catalyst) or through biological activity
within the body. For example *nitrate* can be converted to the
more dangerous *nitrosamine* in the mammalian gut. Other substances act as cancer promoters, i.e. they are not carcinogenic
themselves but can stimulate tumours to grow which have already
been caused by carcinogens. Some carcinogens only promote
cancer in the offspring of the person exposed.

The fact that something is carcinogenic always means that it should be treated very carefully, but not necessarily that it should be banned altogether. Indeed, there is a school of thought which says that virtually anything can be carcinogenic if you have enough of it, simply because of the imbalance it causes the body. Very few of the substances listed below are as dangerous as smoking cigarettes.

The ways in which we test for carcinogenicity are described in the introductory chapters of this book. Some of the wider implications of hazardous substances in the environment are discussed in the afterword. This section simply lists some well known, and a few less well known, carcinogens which you are likely to encounter in everyday life. *It shouldn't necessarily be taken as a list of things to be avoided at all costs.* Nor does it make any attempt to be complete – that would take a book on its own.

Individual entries give more information about levels of risk, where these are known, and ways of minimizing hazards. Many substances in everyday use have not been carefully evaluated with respect to their cancer hazards. Many of those which are listed here remain 'suspected' – i.e. unproven – carcinogens, or those which are carcinogenic in animal studies but which may not necessarily be so for humans.

The whole subject is very, very complicated. But the list does show a small selection of the wide range of materials which do have some known or suspected cancer risks attached to them.

Substance	Source
Aflatoxins	found with some food moulds
Aminotriazole	herbicide
Arsenic trioxide	yacht anti-fouling paint
Asbestos	building material
Benomyl	fungicide
Benzene	*solvent*
Bitumen	sealant, waterproofing
Calcium plumbate	used in some priming paints
Captan	fungicide
Carbaryl	insecticide
Carbendazim	fungicide
Carbon tetrachloride	*solvent*
Chlordane	*pesticide*
Coal smoke	*burning of coal*
Copper sulphate	fungicide

Creosote	*wood preservative*
Cypermethrin	wood preservative and pesticide
2,4–D	herbicide
1,2–dichloromethane	additive to leaded petrol
Dichloromethane	additive in paint removers
Dichlorvos	insecticide in flea collars
Dieldrin	insecticide
Diethylene glycol	stain-removing solvent
Diethylstilbesterol	artificial *hormone*
Dinocap	fungicide
Dioxin	toxic by-product
Formaldehyde	found in furniture, chipboard and so on
Food additives (some)	
Hydrocarbons	pollutants from *road vehicles* and so on
Limonene	lemon scent in air fresheners
Malathion	insecticide
Naphthelene	used in mothballs
Nicotine	insecticide
Nitrosamine	break-down product of nitrate
Pentachlorphenol	insecticide and wood preservative
Perchloroethylene	*dry-cleaning* constituent
Permethrin	synthetic pyrethroid insecticide
Petroleum	
Plasticizers (some)	used in plastics
Polychlorinated biphenyls (PCBs)	by-products from industrial processes
Polycyclic Aromatic Compounds (PAHs)	compounds in tar
Polyurethane	paints
Polyurethane foam	insulation
Radiation	various sources and wavelengths
Radon	released from uranium
Saccharin	artificial sweetener
Selenium sulphide	used in some anti-dandruff shampoo
Solvents (some)	
Sunlight	
2,4,5–T	herbicide
Tar products	widely used in building materials
Tecnazene	fungicide used especially on potatoes
Titanium oxide	white pigment used in some paints

Tobacco	cigarettes, cigars
Trichloroethylene	solvent known as 'trike'
Turpentine	solvent
White spirit	solvent
Zinc chromate	used in some metal priming paints
Zineb	yacht anti-fouling paint

Assessment

This list shows some clear *types* of substances to be wary of – pesticides, solvents, food additives and so on. But beware, more is known about these areas in part because they have been studied more carefully. And not *all* pesticides and solvents are carcinogenic.

Further Information

A fairly heretical view of cancer, which assumes that industrial pollutants play a much more important part than hitherto suspected, is put forward in *The Politics of Cancer* by Samuel Epstein (Sierra Club Books). A new and extensively revised version of this book was produced by Samuel Epstein, Lesley Doyle and six others, aimed at a British audience: *Cancer in Britain: The Politics of Prevention* (Pluto Press, now out of print).

DIOXINS

The Issue

Dioxins are arguably the most toxic substances known. They are released by accident through a number of industrial processes and contaminate the general environment, and also specific products such as certain types of paper and pesticides.

The Facts

In the late nineteenth century, workers in factories making chlorine gas were seen to be suffering high levels of various ailments, including *chloracne*, a particularly serious skin complaint. It was known that something released during the making of chlorine gas was to blame, but dioxins were only finally identified as the culprits in the 1950s.

'Dioxin' is used loosely as a general name for a group of about 75 closely related chlorinated compounds containing two oxygen atoms and two benzene rings. There is a great deal of variation in toxicity between different dioxins. The most dangerous form, which is also one of the most common, is called 2,3,7,8–tetra-chlorodibenzo-p-dioxin, almost always shortened to 2,3,7,8–TCDD, or just TCDD. These substances are extremely toxic. Many are powerful carcinogens and mutagens, cause damage to the immune system and have a range of other effects, including chloracne.

They are stable so that, once in the environment, they can remain active for a long period of time. Dioxins are easily taken up by living organisms, and accumulate, especially in the fatty tissues. Dioxin effects are difficult to identify because there is often a long latency period and indeterminate symptoms.

Dioxins are never manufactured deliberately, but are formed accidentally as by-products of a range of chemical processes. Amongst the common sources are:

- *Waste incineration* (see page 243) and the storage of toxic wastes, especially hospital waste incinerators;
- Pulp and paper mill processes, which contaminate the environment and also leave residues on paper products, including tampons, nappies and toilet paper. This is a particular environmental problem for parts of the United States, Canada, Sweden, Finland and other countries with major pulp-milling operations but dioxins can be found in any paper bleached with chlorine;
- Some pesticides, including *2,4,5–T* and *pentachlorphenol* can be contaminated with dioxin;
- Chemical, plastic and pharmaceutical factories;
- Human breast milk, where levels exceed World Health Organization limits in many parts of the world;

Waste incineration is probably the most important source. In Sweden, where there is major dioxin pollution from paper mills, waste incineration was still the largest single source, accounting for about half the known emissions to the environment.

There have been a number of well-reported scandals concerning dioxins. By far the most infamous is their use in Agent Orange, a defoliant sprayed by American troops during the Vietnam War to destroy cover for the North Vietnamese Liberation Army. Over a period of around twenty years, some 10 per cent of Vietnam was sprayed with Agent Orange, a potent mixture of chemicals made up principally of 2,4,5–T and 2,4–D, heavily contaminated with dioxins.

Abnormally high rates of cancer amongst US troops using Agent Orange, and birth defects in their children, eventually alerted the authorities to what was happening, but even so the mixture continued to be used for years after the facts linking it to health effects became uncontestable. 18,000 ex-troops sued Dow Chemicals, the manufacturers of Agent Orange, in the biggest civil action in history. Although they won record damages, the sums for individuals remained small. Far more damage occurred to the Vietnamese people, who still experience appallingly high rates of birth malformations in the areas most heavily 'treated'. They have received no compensation.

There have been a number of other scandals. In 1976, there was an explosion at a Monsanto chemical factory near Seveso, Italy. Large amounts of dioxin escaped into the atmosphere. In the following months and years, there was a spate of rare illnesses, including chloracne. Several people died and there were unusually high numbers of miscarriages and birth abnormalities. Although the Italian authorities never admitted that there were links between this and the accident, the evidence was sufficiently strong to stimulate new legislation from the European Commission, known as the 'Seveso Directive', about tackling major chemical disasters.

There has also been a persistent debate about dioxins in 2,4,5–T in Britain. The Transport and General Workers Union in Britain organized a campaign to drop the use of 2,4,5–T, which gained support from many local councils, large companies and nationalized industries. The government have consistently refused to ban 2,4,5–T and the argument took a political turn when it was revealed that Margaret Thatcher's husband, Denis, was involved in a firm manufacturing the herbicide. 2,4,5–T has been prohibited in many countries, including the United States, Sweden and Norway.

Dioxins are known to be poisonous. The debate today concentrates on whether the amounts released from *waste treatment plants* (see page 243), paper mills and the amount contained in paper are sufficiently dangerous to worry about. In the United States, the Environmental Protection Agency has judged that TCDD causes an unacceptable human cancer risk at one part per trillion. Many experts judge that there is no no-effect level for dioxins, i.e. they increase risk at whatever concentration they are found. Others, including the British government, believe that the risks at the lowest concentrations are too small to worry about in practical terms.

The impact of dioxins in paper has had considerable publicity in recent years. Studies in the United States and Sweden have shown that dioxins can migrate from paper packaging into food-stuffs, especially if this has a high fat content like milk or cream. This means that dioxins in paper-based milk cartons could migrate into the milk. The US EPA has measured TCDD concentrations in paper as between 0.8–400 parts per trillion. The EPA have estimated that risks of cancer at normal levels of dioxins in paper are 1.4 per million for regular users of paper towels; 1.8 per million for *tampon* (see page 241) users; and 1.5 per million for people using disposable nappies.

Another contentious issue has been the discovery that human breast milk is likely to contain detectable traces of dioxins. Research carried out by the World Health Organization in Sweden, Norway, Belgium, Denmark, Germany, the United States and Vietnam found that babies picked up dioxins through breast milk and dioxins could also be passed directly to the foetus through the placenta. Concentrations are very low and the WHO has concluded that there is no cause for concern, but again this depends on which of the various studies are used to make risk assessments.

Assessment
Despite the huge amount of work carried out on dioxins, it is still extremely difficult to make anything but very broad statements about risk levels. The EPA studies have concluded that risks from everyday use in paper products and so on remain very small, but not non-existent. (Other researchers would disagree about the latter point.) There is certainly reason to be concerned about high levels of contamination, through waste incineration or deliberate or accidental contamination. It is also worrying that dioxins have built up so widely in the environment that they are likely to turn up routinely in the human body.

Minimizing Risk
Unfortunately, there's not much you can do about dioxins in the wider environment, beyond protesting against the weak laws that allow this kind of contamination to continue. However, there are a few particular situations which might increase your

exposure – and these can be avoided.

- Pesticides: although makers of 2,4,5–T claim that it isn't contaminated with dioxins any more, there are plenty of safer alternatives available, if you need to use herbicides at all. (2,4,5–T has itself been banned in some countries because of alleged health risks);
- Paper products: unbleached paper products are becoming more widely available as people become worried about dioxins. There are also new bleaching methods being introduced which do not use chlorine, and thus avoid dioxin formation. Other ways of reducing exposure are described in the section on *tampons* (see page 241);
- See also the section on *waste treatment plants* (see page 243).

A great deal of work has been carried out on the dioxin issue in Sweden, where fears about levels in paper products are gradually changing practices in pulp mills. Several useful publications are available in English from the National Swedish Environmental Protection Board (Solna, Sweden). See, for example, *Biological Effects of Bleached Pulp Mill Effluents* and *Dioxins: A Program for Research and Action*.

In Britain, the *Women's Environmental Network* (377 City Road, London EC1) has carried out a great deal of research on this issue. See the section on *tampons* for further information (see page 242).

ELECTROMAGNETIC RADIATION

The Issue
It is now thought that Extremely Low Frequency (ELF) radiation, from power lines, microwave ovens and so on, can be damaging to the health. This non-ionizing radiation has received far less public attention than ionizing radiation from nuclear power stations, *X-rays* and *sunbathing*.

The Facts
A few years ago, someone noticed that staff at the American embassy in Moscow were suffering from an unusually high rate of blood disorders. The suspected cause, carefully leaked to the Western press, was high levels of Extremely Low Frequency

(ELF) radiation occurring in the embassy which, according to the US government, was occurring as a result of a bugging operation by the KGB.

Since then, deliberate or accidental exposure to ELF radiation has become one of the most talked about new environmental dangers internationally, especially in the USA. In 1986, the Houston Power and Lighting Company was fined $25 million costs because they erected a 345 kV power line within 60 metres of three schools. At around the same time, the women peace protesters at Greenham Common in England were complaining that ELF radiation was being deliberately directed at their bender camps by the US government in the hope of forcing them to leave.

ELF radiation is emitted from a range of modern electrical appliances including:

- High voltage electricity power lines;
- CB radios;
- Television and radios;
- Electric blankets;
- Radar dishes;
- Some security systems;
- *Microwave ovens*;
- *Visual display units*;
- Lasers, etc.

Of these, the best known source is electric power cables. The amount of ELF in the general environment has increased enormously over the last few years. The average American now receives two hundred times more than the 'background level' of electromagnetic radiation experienced 100 years ago.

The precise impacts on health are not understood, or agreed. Edward Goldsmith and Nicholas Hildyard compiled information on risks from various sources in their book *The Earth Report*, and presented known or suspected risks from a 400 kW power line over a range of distances. They found evidence for a range of effects, varying with distance. For instance, at 30 metres there were reports of allergies, blackouts, concentration loss, epileptic fits, exhaustion, headaches, increased leukocyte count, loss of muscle power and palpitations. Even at 245 meters there was still some evidence of convulsions in allergic subjects, epilepsy and rare eye cancer.

Other evidence is being pieced together. For example, research has shown that people working in electric and magnetic fields had increased risks of developing leukemia. By early 1986 a survey showed that fifteen out of seventeen studies had found increased risks of leukaemia among electrical and electronic workers. Childhood mortality records around Denver, Colorado, found increased incidence of childhood cancer in the vicinity of high current power lines. The New York Power Lines project, funded by the New York State Public Services Commission, looked at a range of cases and found that if the various hypotheses were correct: 'this could mean that 10–15 per cent of all childhood cancer cases are attributable to electric fields'.

Much of this research is still highly speculative of course, and some of the studies quoted above are anecdotal. At the moment, perceptions of risk varies enormously. In the USSR it is apparently regarded very seriously, while in Britain and a number of other European countries the issue is hardly ever discussed. There does seem to be a growing consensus in those countries where it is carefully studied that the hypothesis of a link, especially with cancer, is worth taking seriously.

In June 1990, a leaked document from the US Environmental Protection Agency suggested that experts are increasingly concerned about the effects of ELF on human health. The report, prepared by agency staff, classifies ELF as 'probable human carcinogen'. In addition, *radio frequency radiation*, from cellular and cordless telephones, radio and TV broadcasts and satellite stations, was defined as a 'possible human carcinogen'. But, according to the New York newsletter *Microwave News*, the ELF classification was deleted by the Agency's Office of Health and Environmental Assessment, after discussions with the White House. Despite the deletions, the published report still concludes that magnetic fields from power lines, and perhaps other sources in the home, were a 'possible but not proven cause of cancer in people'. The strongest evidence, according to the report, comes from studies of increased leukaemia incidence in children of men with electrical jobs and increased cancer rates among telephone workers.

If these theories are halfway correct, they have enormous implications for the modern world. They could result in radical restructuring of the power industry for example. But they would also relate directly to many other issues discussed in this book, including microwave ovens and visual display units. At the moment, it is still too early to make much sense out of what is happening, but this is an issue which is likely to develop very quickly over the next few years.

Minimizing Risk
There is not much the individual can do about power cables or large scale sources of ELF. However, people who are concerned should certainly make sure that sources of ELF in the home, such as radios, microwaves, televisions etc, are not left switched on when they are not being used. The old adage that it is bad to sit too close to the TV may well be right after all.

Further Information
Electropollution: How to protect yourself against it by Roger Coghill (Thorsons).

FORMALDEHYDE

The Issue
Formaldehyde is a common ingredient in resins and glues, and turns up in insulation material, furniture and chipboard. It is an irritant, and there are fears that it may also cause cancer.

The Facts
The safety of formaldehyde is one of the many issues where there is still an enormous divergence of opinion, and where feelings run very high.

Formaldehyde (also known as methanal) is a colourless gas with a pungent and unmistakable smell. It can be dissolved in water to make formalin, which is the form in which it is usually used.

Formaldehyde is a component of glues and resins, and is frequently used in chipboard. It is a major component in some kinds of foam *insulation* (see page 108). Formaldehyde gas is also used as a soil sterilant, fumigant fungicide for glasshouses, and disinfectant. It has also been used as an ingredient in some shampoos. Formaldehyde gas is highly toxic if inhaled in large quantities, and is an irritant to the eyes, lungs and, in high concentrations, to skin as well.

The main sources of formaldehyde in the home are:

• *Urea formaldehyde cavity wall insulation*: some types of cavity wall insulation release formaldehyde gas, especially just after installation;

- *Furniture and chipboard* can also release formaldehyde gas.

There are well documented cases of people suffering irritant effects when new furniture is delivered, or just after insulation foam has been injected into cavity walls. Relatively high levels are likely in mobile homes or caravans using large amounts of chipboard in their construction.

A much more serious issue is the suspicion that formaldehyde can cause cancer. Evidence from experiments on rats in the United States suggests that it may cause lung cancer, although corresponding epidemiological studies on human populations have been inconclusive. Two influential American bodies, the National Institute for Occupational Safety and Health (NIOSH) and the Environmental Protection Agency (EPA) list formaldehyde as a 'potential carcinogen' but this is strongly refuted by some other bodies, including the British government.

The World Health Organization, in a special study, states:

Epidemiological studies do not clearly indicate carcinogenicity in humans, but because of the poor power of the best studies to date, the possibility that formaldehyde is a human carcinogen cannot be excluded.

This conclusion is strongly refuted by some other bodies, including the British government, who have given it the all clear.

The WHO study also lists a set of problems which require further research, including allergies, changes in immunological parameters, pulmonary functions, physical and mental development in children, carcinogenicity and interaction with other factors, including particles, fibres, aero-allergens and nitrogen oxide.

The balance of evidence on health effects has led the United States authorities to restrict the use of cavity wall foam insulation based on formaldehyde, and to consider further restrictions on other uses. Apparently, Germany is also considering tighter restrictions. The British Health and Safety Executive has re-examined the evidence and decided that there are not sufficient grounds of withdrawal, and this stance has been backed by some very forthright statements from ministers in parliament. They argue that the US position is alarmist and, in any case, based on very different styles of building, so that risks of irritation from formaldehyde in American timber-frame houses are greater than in houses conforming to British building regulations.

This illustrates one of the major bones of contention, namely how much formaldehyde you are actually likely to be exposed to from insulation or furniture. An industry spokeswoman once told me that there was higher concentrations of formaldehyde in the average bread bin due to emissions from bread than in most houses that have been insulated. On the other hand, relatively high concentrations have been recorded on occasion.

Assessment
It is probably still too early to say. The industry has such a powerful and well-organized lobby in Britain that rational discussion has been hampered over the last few years. It is certainly worth taking the simple precautions listed below.

Minimizing Risk
Try to avoid any products containing formaldehyde if you suspect or know that you are sensitive to them. In addition:

- Ensure adequate ventilation if products likely to release formaldehyde are in your home. This means that it's best to have cavity wall insulation containing formaldehyde injected in the warm days of the summer, when you can keep the house well aerated for a few weeks. Watch for any signs of irritation, headaches and so on in people living in the house;
- Seal all chipboard, plywood and furniture likely to contain, and release, formaldehyde gas with varnish or paint;
- Anyone using formaldehyde as a sterilant should take very great care as this is admitted to be a very hazardous operation. The British Crop Protection Council advises: 'After fumigation allow at least three days before entering structure or planting a crop.' After soil treatment, 3–6 weeks must be left before planting anything else.

Further Information
For general information on formaldehyde in buildings, contact the *Hazards Bulletin* in Sheffield (3 Surrey Place, Sheffield S1 2LP). See also *Indoor Pollutants* from the National Academy of Sciences, Washington DC (1981), and *Indoor Air Quality:*

Radon & Formaldehyde from The World Health Organization Office in Copenhagen.

GENETIC ENGINEERING

The Issue
Genetic engineering refers to the practice of altering the genetic make-up of natural plants and animals in order to produce something substantially different. Genetic engineering is still a new technology, but is attracting huge amounts of investment from large corporations, and is being developed very quickly. Many environmentalists view it as potentially disastrous if not properly controlled.

The Facts
Successful genetic engineering is less than twenty years old. In 1973, researchers at the American universities of Stanford and California performed the first successful experiment in 'recombinant DNA', which means the direct manipulation of genetic material (the DNA) within a living organism. In a rather over-simplified explanation, scientists practising recombinant DNA, more usually known as genetic engineering, are working directly with the chromosomes that carry genetic (hereditary) information. They can add pieces from the chromosomes of other individuals, or other species, take bits away and tamper in a variety of other ways. Once the chromosome is altered, the characteristics it controls will, with luck, be passed on to descendants of the manipulated plant or animal.

Human attempts to alter other living creatures are nothing new. Plant and animal breeding was practised by the earliest civilizations. By successively crossing species with the characteristics the farmer wants to emphasize, we have created all our modern farm animals and crop plants. The pig bears little resemblance to the lean and hairy wild boar from which it has been bred. However, genetic engineering allows us to contemplate alterations to living creatures on an infinitely greater scale than has been possible until now.

Once the technique has been mastered, scientists are able to perform extraordinary feats of manipulation. Enthusiasts talk of huge fruit and vegetables, programmed to grow in cubical shapes so that they can easily be transported from the factory to the din-

ner table; of cereals which collect their own nitrogen requirements from the air; of cows which produce three times as much
milk as they do now; of enzymes which can dissolve ores from
rock and replace miners; and so on and so on.

Genetic engineering, which is also known as *biotechnology*, has
two main groups of supporters. Some people believe that it is a
technology which can transform human society, feed all the malnourished people in the world, cure all illnesses and usher in a
new golden age. Added to these are many companies who see the
potential for vast profits.

So far, genetic engineering hasn't really impinged on our
everyday lives. But it probably will, very soon. This section outlines some of the fears that environmentalists, and others, have
expressed about the further use and uncontrolled development of
genetic engineering. A number of objections have been raised.
The first, and most obvious, is that while it may well be possible
to breed things which could be useful, there are also ample
opportunities to create monsters. Mary Shelley's book
Frankenstein could well have been written as a warning about
genetic engineering! By 'monsters' I don't mean hulking brutes
from the science fiction movies so much as new viruses, bacteria
and other pathogens which can harm us, our crops or the natural
world.

These fears are not alarmist. It is inevitable that genetically
modified organisms (GMOs) will escape, or be deliberately
released, from a laboratory into the environment. 'Once it has
escaped, there will be no way to recapture it,' according to
Professor Robert Sinsheimer, Chairman of the Biology Division
of the California Institute of Technology, 'and so we have the
great potential for a major calamity.' You can't go out and recapture a virus with a butterfly net.

There are other worries as well. The increasing privatization
of research has meant that the direction of genetic engineering
research has not been necessarily towards what is best for the
general human good so much as what is best for the corporations running the laboratories. So, for example, large chemical
companies are now developing strains of crop plants which are
resistant to *herbicides* produced by the company; not resistant to
pests or to disease, but to a chemical which can then kill *other*
species. The process by which major agribusiness companies
breed crops which then require other products from the company, such as extra fertilizers, pesticides and so on, can be

refined and accelerated with biotechnogy.

In addition, ill-used biotechnology may not be good for the planet as a whole. British people were shocked recently by photographs of grossly fat and distorted pigs bred through genetic manipulation; good news for the pig farmer's profits perhaps, but consigning the pig to a life of misery. (Many farmers reject such practices.) Super-breeds of plants and animals could outcompete natural species and destabilize ecosystems. Everyone working or researching the field agrees that we are playing high stakes with genetic enginering.

So far, attempts at control are not working very well. In 1974, the US biologist Paul Berg, who was one of the original researchers working on recombinant DNA, called for a moratorium on genetic engineering of carcinogenic, mutagenic or teratogenic organisms. It did little to halt research. In 1980, the pro-biotechnology lobby in the United States won the right for companies to patent novel forms of life; this means that if a company actually does develop something useful they own it and can charge what they want for it.

At the moment, there is supposed to be an international moratorium on the release of genetically modified organisms into the environment. Again, it doesn't mean much in practice. The British government has been secretly allowing the sales of milk containing *bovine somatotropin* (see page 38), a genetically modified growth hormone. In Australia, meat from 53 genetically-engineered pigs has been eaten by people without labelling, after the pigs were sold from laboratories at the University of Adelaide. The University authorities apparently knew that they were in breach of the code, but did not even inform their own biosafety committee.

Over the next few years, there will be an increasing debate about the potential health and environmental effects of genetic engineering, biotechnology and the release of novel forms of life. Remember when you are reading the literature from the companies just how much money is at stake!

Further Information

There are a number of excellent books on the subject. Unfortunately, some of the best are already out of date. *The Gene Business* by Edward Yoxen, published by Pan Books in Britain in 1983 gives a clear and readable background, with interesting insights into the politics involved. *Algeny* by Jeremy Rifkind is readable and passionate; Rifkind is the best known campaigner on these issues, working from the United States. Another readable account is *Non-Rich Life: A Genetic Engineering Paper* by Jeremy Cherfas (Blackwell 1982).

For further information, The Genetic Forum (3-4 St Andrews Hill, London EC4V 5BY, 071-236-8373) is an umbrella grouping of several different organizations interested in these issues.

MUTAGENS AND TERATOGENS

The Issue
Mutagens and teratogens are substances which increase the risk of genetic mutations and birth defects.

Mutagens are substances which cause mutations in the *off-spring* of the affected individual. A *mutation* is a change in the chemical structure of the genetic material, which can in turn cause changes in the body. While a few mutations can be advantageous (and are the cause of much evolution) the vast majority are harmful. Mutagens are also often carcinogens as well.

Teratogens are substances which cause deformities in the off-spring of parents exposed to them. Many, but not all, teratogens are also mutagens, i.e. deformities are often caused by mutations in genetic material.

Testing for mutagens and teratogens is still by no means well understood. It is often assumed that these only work through the mother, but in fact mutagenic and teratogenic problems can often pass to a child via the father as well.

The Facts
The problems of assessing risks to health from mutagens and teratogens are much the same as for substances which increase the risk of cancer, and are further discussed in the section on *carcinogens* (see page 208); read the introductory chapter for cautions and qualifications. In this section, a checklist of substances known or suspected of being mutagenic or teratogenic is given.

Substance	Source
Allethrin	insecticide, often used indoors
Aminotriazole	herbicide

Atrazine	herbicide, occurs as residue in water
Benomyl	fungicide
Calomel	fungicide
Captan	fungicide
Carbaryl	insecticide
Carbendazim	fungicide
Chlordane	pesticide
Copper sulphate	fungicide
2,4–D	herbicide
Demeton-s-methyl	insecticide
Diazinon	insecticide used in cat flea collars
Dichlofluanid	used in timber treatment
Dichlorvos	insecticide, also used in flea collars
Dieldrin	insecticide and timber treatment
Dioxin	by-product of some pesticides, waste burning, chemical production and so on
Diquat	herbicide
Fenchlorvos	insecticide (including professional use in households)
Fenoprop	lawn weedkiller
Ferrous sulphate	moss killer
Malathion	garden insecticide
MCPA	lawn treatments
Mecoprop	lawn treatments
Nicotine	insecticide
Perchloroethylene	dry cleaning fluid
Pirimiphos methyl	organophosphate insecticide
Plasticizers (some)	used in plastics
Propachlor	weedkiller
Radiation	
Simazine	weedkiller
Sodium hydroxide	caustic soda
Sodium hypochlorite	bleach
Solvents (some)	
Sunlight	
2,4,5–T	herbicide
Tobacco	
Toluene	solvent
Trichloroethylene	solvent
Zineb	yacht anti-fouling paint and insecticide
Ziram	yacht anti-fouling paint

Further Information
See the section in *carcinogens* for details of background
information and contacts (see page 208)

NITRATE

The Issue
Nitrate levels in both drinking water and food are rising as a result
of changes in farming methods. There are continuing questions
about the effects of nitrate on human health.

The Facts
Nitrate occurs naturally in the environment, and is essential for
crop growth, but modern, chemical, farming methods have
increased the amounts found. Excess nitrate comes from a number of sources, including:

- Soluble nitrate fertilizers added to crops and pasture;
- Nitrate released whenever land is ploughed up (so that intensive arable farming and ploughing and reseeding pasture
  increases the amount of mobile nitrate in the environment;
- Nitrate deliberately added to meat as a preservative, where it is
  used to prevent botulism.

Excess nitrate causes a number of environmental problems,
including water pollution through algal blooms (where it acts
with phosphate which is another important pollutant), and – in
different forms – as an air pollutant.

Some of this nitrate ends up in our diet. Nitrate levels in drinking water have been increasing fairly quickly in some areas. In
Britain, for example, about 1.6 million people now drink water
with nitrate levels higher than the legal maximum set by the
European Community. Certain vegetables can build up high
nitrate levels as well, especially if they are grown intensively,
under glass, in the winter period; both lettuce and spinach can
sometimes contain high nitrate levels, although concentrations
are extremely variable. The nitrate in meat is generally quite a
small proportion of the total, averaging about 15 per cent of the
proportion in solid food.

The percentage of dietary nitrates from water, and thus the
total intake, depends very much on where you live, and the time

of year. In intensively farmed areas nitrate levels in the water are highest in the winter, when the ground is bare and fertilizer is washed quickly away to streams and downwards into ground-water aquifers. (Although in certain conditions it can apparently take years, or even decades, for nitrate to percolate down to aquifers, so that pollution may continue to get worse for a long time to come.)

There are two main health issues. High nitrate levels in water, if used to feed infants under three months old, can result in the development of *methaemoglobinaemia*, commonly known as *Blue Baby Syndrome*, a potentially fatal condition. At the moment, this isn't at all widespread; there have only been about 2,000 cases worldwide in the last few decades. Risks are increased by other factors, including poor sanitation and malnutrition. The last case in Britain was in 1972, and only one known fatality has occurred in the UK.

A potentially much more serious issue is the possibility of links between nitrate and stomach cancer. One of the breakdown products of nitrate which can be formed in the human body, nitrosamine, is intensely carcinogenic in well over thirty species of laboratory animals, including primates. But epidemiological tests on human populations have failed to find clear evidence of links with stomach cancer, believed to be the most likely type of cancer to be formed.

However, there are lots of complications. Most people get the bulk of their nitrates from vegetables, but this is also an important source of vitamin C, which is thought to *protect* against stomach cancer. Perhaps the two factor are, generally speaking, cancelling each other out. Stomach cancer is declining in the developed world. It is thought that this is mainly because of increased food hygiene and the elimination of many carcinogenic moulds (see *aflatoxins* page 29). Any slight increases caused by nitrates may not be detected in the more general trends. In addition, the increase in nitrate levels is still quite new, so effects may not have shown up in human populations as yet.

Assessment

Theoretically a dangerous substance, but at present research evidence on human populations does not bear this out. Watch out for further information though; the researchers who work on these issues tend to be a lot more cautious than the publicity people at the fertilizer companies!

Minimizing Risks

If you want to reduce nitrate levels in your diet, there are a number of options:

- Don't buy 'at risk' vegetables out of season, in the winter, when nitrate levels are likely to be high;
- Use a water filter or buy mineral water (but see *water filters* page 72) and *mineral water* (see page 74) for qualifications to this advice);
- Buy certified organic meat, which has not been preserved with high levels of nitrates;
- In some countries, including Britain, local authorities supply bottled water to nursing mothers in areas with high nitrate levels. Anyone worried could buy mineral water, boil it, and use this instead.

Nitrate is found in the plant's cell walls and cannot be eliminated by washing or by cooking. Boiling water doesn't help either; if anything it concentrates the nitrate even more. Very few water filters are effective in reducing nitrate levels.

There is a much clearer picture regarding use of nitrate in meat, and urgent research is required to find safer alternatives. This has been accepted for some time, and the US National Academy of Sciences published an entire report on the subject, with a recommendation that nitrate be replaced with something else, perhaps ascorbic acid, back in 1980.

Several European countries, including Germany, Switzerland and Austria, have introduced controls on nitrate levels in vegetables. The Soil Association called on the British government to do the same in December 1989.

What Governments Could Do

There is now an overwhelming consensus that we have to reduce nitrate levels in water, but the methods being suggested are still nowhere near strong enough. Organic farming methods offer a chance radically to reduce nitrate leaching, through eliminating soluble nitrate fertilizers and careful management of wastes to reduce leaching. But *any* agriculture will increase nitrate leaching into water to some extent.

Further Information

See *Nitrates* by Nigel Dudley (Green Print) for a general

overview. There are many books and reports available on this issue, both from the fertilizer industry and national and international bodies, including the National Academy of Sciences in the USA, and the Royal Society in Britain.

NOISE POLLUTION

The Issue
Has the use of stereo headphones, ghetto blasters and loud discotheques really ruined the hearing of today's (and yesterday's) youth? Or is it just a baseless rumour put about by old fogies too boring to enjoy themselves!?

The Facts
Unfortunately it's all too true. Survey after survey has found that people constantly listening to excessively loud music in any form suffer exactly the same kinds of problems as people working in a very noisy environment; i.e. premature deafness and *tinnitus* – or constant ringing in the ears.

These are two rather different effects. Premature deafness means that the clarity of your hearing is reduced. Common manifestations of this are:

- Not being able to distinguish high pitches very well so that, for example, you can't here most bird song;
- Not being able to distinguish different sounds occurring together, such as happens when several people are talking together at a party.

Tinnitus is a very common complaint that many people may have without even being aware of it. It is a (usually fairly tinny) ringing in the ears. In most cases you will only really recognize this when there is complete silence, such as in the middle of the night or in a country area. I pick it up in myself slightly whenever I am alone in the mountains and sit down quietly – which shows just how noisy the modern world is the rest of the time! However, for people with serious tinnitus it is a constant and extremely annoying irritation that you have to learn to live with night and day. Pete Townsend, of the Who, suffers this after years of playing loud rock music. It is at present virtually incurable and very distressing in extreme forms.

Minimizing Risks

Although any loud noise is dangerous, for most people reading this book the greatest risks come from listening to loud music, working with noisy machinery or, perhaps, through their work in noisy factory or tool room.

Perhaps one area worth expanding is that of risks from stereo headphones, because these give people the chance to carry the potential for aural self-abuse about with them wherever they go. Everyone must have sat by people on trains or buses and heard the insistent tinny underlying beat of the music coming through. The chances are that this means the 'intentional listener' is hearing the music too loudly for safety.

Unfortunately, there aren't any hard and fast rules about the maximum level music should be played. If your ears ring or hurt after listening to music, then it is almost certainly too loud. If you keep the music to the top of the volume setting, it is probably too loud as well. In general, it is worth trying to train yourself to listen to music being played more softly. On all but the most expensive machines there is considerable distortion at the top of the scale anyway.

There are particular dangers for children, who will often turn the music up as loud as they can just to experiment. Stereo headphones made especially for kids, which don't go too loud, are available and a good investment unless you are convinced that children in your care will not unintentionally damage themselves. Young children are also more likely to suffer hearing damage from loud music played at discos or parties.

For people at work, there are now detailed regulations governing noise level in most countries. However, these are by no means always observed. If you think that you are suffering from over-loud music or other sound, contact the management or the trade union representative.

Further Information

In Britain, the Noise Abatement Society (PO Box 8, Bromley, Kent BR2 0UH, 081-460-3146) can offer leaflets and advice.

POWER STATIONS

The Issue

There has been a lot of concern about the environmental effects of power stations. At present, we know a great deal about the effects of coal-fired power stations on freshwater ecosystems and, to a lesser extent, on forests and other land habitats, but very little about overall health effects on people. We know a little more about nuclear power stations. This section provides an overview and directs the reader to other sections of the book containing more information.

The Facts
Power stations, which generate electricity, are some of the great energy-consumers of our time, and also some of the most important polluters. From the perspective of direct effects on human health, the main issues are:

- The effects of pollutants (mainly sulphur and nitrogen oxides) from coal-fired power stations.
- The effects of both low-level radiation, and occasional accidental or catastrophic releases of radiation, from nuclear power stations. This is discussed under *radiation* (see below) which looks at other artificial sources of radiation as well;
- The effects of *electricity power lines* in releasing microwave radiation into the environment;
- The wider effects of electricity generation on the *greenhouse effect* and changes in global climate, and on acidification of freshwaters, which can in turn have a number of health effects.

Further Information
National offices of Friends of the Earth and Greenpeace should be able to provide information on the energy supply industry and its polluting effects.

RADIATION

The Issue
Certain wavelengths of radiation – from a variety of sources, including sunlight, *X-rays* (see page 250), natural releases of *radon* (see page 138); and from accidents and routine discharges from nuclear power stations – are dangerous. But quantifying this danger is still extremely difficult, and there is an enormous amount of disagreement between experts.

The Facts

We are constantly being bombarded by radiation, from both natural and artificial sources. Natural radiation comes from the sun, natural rocks, the soil, from a variety of building materials, and from radioactive particles which enter the body.

In industrialized countries and, increasingly, in the Third World as well, we are being exposed to artificial radiation in addition to background levels. The best known source of artificial radiation is nuclear power stations, but we are also exposed to radiation from fallout from nuclear weapons tests, and X-rays used in medicine. People also increase their exposure to natural radiation by *sunbathing* (see page 166) using *sun beds* (see page 167) constructing buildings in ways which increase exposure to *radon* gases (see page 138), and through their work. For example, airline crews can virtually double their annual exposure to cosmic rays by constantly flying at high altitudes, where the rays are stronger; miners can come into contact with high levels of radon gas. We have also increased exposure to cosmic rays by damaging the *ozone* (see page 157) layer which helps repel the more dangerous forms of cosmic rays.

We know for certain that high levels of certain kinds of radiation are extremely dangerous to health, causing both immediate and long-term health problems. Long-term problems include the formation of cancer and arange of genetic mutations, including many birth defects.

For example, many people living in Hiroshima and Nagasaki at the time when the United States dropped nuclear bombs suffered 'radiation sickness' and died fairly soon afterwards; others died years later from cancer. A similar pattern of immediate, medium-term and long-term health effects appears to be taking place amongst people who were living near the site of the Chernobyl nuclear disaster in the USSR.

Immediate effects of radiation sickness include vomiting, loss of hair, bleeding (especially at the gums) and, in a fairly short time, death. Images of people suffering from the effects of radiation have become part of the mythology of the twentieth century, and the discussions about the risks of nuclear accident or nuclear war.

However, the effects of low level radiation are much, much more controversial. In the early days of working with radiation, people were very unaware of hazards to health. Marie Curie, the 'discoverer' of radiation, died of cancer almost certainly brought on by what in retrospect we can see was her reckless handling of highly radioactive material.

For a long time afterwards, it was believed that there was a 'safe limit' of radiation below which people wouldn't come to harm. British soldiers watching atomic bomb tests in Australia were told to simply turn their backs to shield their eyes from the glare; large numbers subsequently developed cancer and other health problems, and sued the British government. (The aboriginal people living in the area also suffered badly, although their plight has received far less publicity.) Looking at early literature about civil defence at times of nuclear war shows how hopelessly the authorities had underestimated levels of risk. Workers in nuclear power stations, naval shipyards building nuclear submarines, X-ray operators and others have, for years, been exposed to levels of radiation which are now believed to be unsafe.

Uranium miners have suffered even more badly, usually because they have come from groups without a powerful political voice; they include the Navajo Indians in the United States, and the black people of Namibia and the aboriginals of Australia.

Today, it is accepted that far lower levels of radiation are dangerous than was believed to be the case in the past, but there is still little agreement about the overall levels of risk from low-level radiation. The International Commission on Radiological Protection (ICRP) bases its assessment of radiation damage on the premise that any dose of radiation – however small – is potentially harmful.

The ICRP also bases its calculations on a straight line relationship between the radiation dose and the probability of developing cancer, setting the upper limits on levels of cancer seen in 'survivors' of Hiroshima. Such an approach has attracted criticism, for two reasons.

First, there is no clear agreement about the straight line relationship between dose and effect. Some scientists believe that high doses are proportionately far more dangerous, and that low-level radiation is, for all practical purposes, harmless. Others think the reverse; that low-level radiation is particularly dangerous in that it stimulates the body's repair mechanism, which thereafter operates imperfectly and increases risk from later exposure. Second, there is disagreement about the use of the Hiroshima statistics. Some researchers believe that this leads to a serious underestimate of risks, because only the healthiest people survived the *initial* exposure to radiation in the nuclear explosions.

This means that there would be an artificially healthy and resistant population sample, who might also be particularly resistant to developing cancer later in life. This would lead to an underestimate of risk at the top end of exposure, which would then be applied to all other exposures as well.

It is also accepted that some sources of radioactivity can become concentrated in the food chain, in the same way as persistent pesticides, heavy metals and some other pollutants. Fish and shellfish caught near outflows from nuclear power stations have sometimes been found to contain higher than average levels of radiation. This can be particularly hazardous if it is in a form which can be stored in the body, for example as a particle of plutonium which can become lodged in the lung or elsewhere.

The problem with establishing the impact of low-level radiation is that – as with other carcinogens discussed in this book – it is very difficult to count the bodies. The nuclear industry has fought a long and shameful battle against any admission that its industry is dangerous. Questions quickly become political. Sorting out the real facts has taken a long time. Simple correlations between incidences of cancer and, for example, proximity of nuclear power stations, are not enough to prove a connection between the two. There is still much research to be done.

Elsewhere in the book are sections on the radiation risks of *radon* (see page 138), *sunbathing* (see page 166), *sun beds* (see page 167) and *X-rays* (see page 250). These are among the most important sources of radiation for people in everyday life. There is also a section on *irradiation* of food (see page 49). However, the best known source is radiation is from *nuclear power stations.*

The nuclear industry has long argued that nuclear power stations are perfectly safe – 'atoms for peace' as President Eisenhower said. People arguing against nuclear power on safety grounds have been accused of scare-mongering, communist agitation, anarchy and much else besides. However, research at nuclear power stations, especially in Britain, has established two important principles.

First, that workers in nuclear power stations have suffered higher than average levels of some cancers. British Nuclear Fuels Limited have accepted tacit responsibility for some incidents, by paying compensation to workers and their widows, without ever publicly admitting that radiation was to blame. (And once again, we must remember that it is impossible to *prove* that any particular carcinogen is responsible for producing any particular cancer.)

Deaths to date among workers at the Hanford nuclear establishment in the United States indicate that cancer rates for workers exposed to the highest radiation levels are 25 per cent higher than average. The investigation which compiled these statistics was first completed in 1977 and though an attempt was made to suppress it, it was published and has subsequently been confirmed by other independent studies.

Secondly, it has been shown that there are higher than average clusters of leukaemia in children living around nuclear power stations. This doesn't *prove* that radiation from nuclear power stations was at fault. But the Committee on Medical Aspects of Radiation in the Environment, which acted for the British government in investigating child cancers around Sellafield and Dounreay nuclear reprocessing plants, concluded that they were probably to blame for the leukaemia cluster. Professor Martin Bobrow commented: 'You can never rule out chance, but we've found an excess of cancer cases at the only two nuclear reprocessing plants in Britain.' The reasons for this are still not clear: routine discharges are theoretically not high enough above background levels to account for the increase. Possibilities include the chance that discharges find 'novel pathways' in the environment, thus increasing exposure, or that parents working at the plant pass on some 'preconception effect'.

Acceptable doses for both nuclear industry workers and for the general public have been lowered several times over the past few decades. According to researchers such as Dr Alice Stewart of the Cancer Registry, Queen Elizabeth Hospital in Birmingham, England, these are still too high for safety.

Assessment

There are risks from low-level radiation from nuclear power stations. These are probably fairly low compared to other risks; for example Dr Alice Stewart estimates that 70 per cent of childhood cancers are caused by natural background radiation. Others will be caused by *X-rays* and exposure to other, non-radioactive, carcinogens. But people living near a nuclear power station, or reprocessing plant, or somewhere making nuclear reactors, are justified in being worried.

Far more worrying, and the basis of so many of the protests about nuclear power stations, is the risk of catastrophic breakdown. The accident at Chernobyl is by no means the most serious that could occur. Farmers in Wales are still unable to sell sheep because of the high radiation levels caused by fallout from Chernobyl, five years after the explosion and many thousands of miles away. The lifestyles of the Lapps of northern Scandinavia has been altered, probably forever. Of course, the risks of radiation in the event of a nuclear war would be altogether catastrophic for the whole of humanity.

Further Information
There has probably been more written about the health effects, politics and history of nuclear power, and of radiation, than any other single topic covered by this book. For a good introduction to issues of radiation risk, try *Nuclear Power* by Walter C. Patterson (Penguin) and *Britain's Nuclear Nightmare* by James Cutler and Rob Edwards (Sphere). For a detailed breakdown of low-level radiation issues, see *Low Level Radiation: Questions and Answers* by Patrick Green, published by Friends of the Earth in Britain and *Radiation Risks: An evaluation* by David Sumner (Tarragon Press). There is also an enormous amount of information published by the nuclear industry.

SOLVENTS

The Issue
'Solvents' are, literally, liquids which dissolve things. Solvents are used as 'carriers' for pesticides, adhesives, cosmetics, paints and so on, and are used directly as, among other things, anti-freeze, paint removers, paint brush cleaners. Most of those used commercially are quick to evaporate (i.e. they are volatile) and many are toxic. There has been a flurry of interest in solvents recently because some are *ozone* depleters (see page 157), but many can also have more direct effects on human health. This section provides an overview of the risks, possible risks, and precautions to take with a range of common solvents.

The Facts
In the following table, solvents are listed alphabetically, the major uses or sources are given, and some of the main effects outlined. It should not be assumed that because something is listed as being found in a particular type of product that it will be found in all such products; for example amyl acetate is found in by no means all types of nail varnish. It depends on where you live as to

how much information about ingredients will be listed on bottles or packets of many items likely to contain solvents.

Solvent	Uses/Sources	Effects
Acetone	grease remover, nail varnish etc	moderately toxic if swallowed or inhaled, and can cause skin irritation
Amyl acetate	lacquer, nail varnish etc	moderately toxic if swallowed or inhaled, and a potential skin irritant
Benzene	found in various petroleum mixtures	acutely toxic, also a chronic poison. It can cause a range of health effects, including blood disorders ranging from anaemia to leukaemia; also fatigue, weakness etc. It is a powerful irritant. One of the components of petrol (gasoline) which make it dangerous to inhale or splash onto the skin
Butanol	general solvent	moderately toxic if inhaled, swallowed etc and can also cause skin irritation

Butatone *see* methyl ethyl ketone

Carbon tetrachloride	grease remover and also found in some fire extinguishers	very toxic if swallowed or inhaled and fairly toxic if absorbed through the skin. An irritant and suspected carcinogen. Has been banned in many countries because of its acute toxicity, but can still be seen used by hand as a cleaning material for engines in some Third World countries. Reasonably

powerful ozone depleter which has, so far, not been covered by any international protocols

Cellulose acetate *see* ethyl glycol acetate

Cellulose solvent *see* 2–ethloxyethanol

Dichloro-methane	paint removers	moderately toxic if inhaled, can cause skin irritation. Suspected carcinogen and teratogen
Diethyl glycol	stain removers	highly toxic if inhaled. Suspected carcinogen. Skin irritant. Can contaminate food if used in cellulose packaging and is no longer used for this purpose in the United States
Ethanol	DIY (Do It Yourself) and cosmetic products	the intoxicating compound in alcoholic drinks. Moderately toxic if swallowed or inhaled and can irritate the skin
2–ethoxy-ethanol	used in cellulose paints	moderately toxic if swallowed or absorbed through the skin
Ethyl acetate	lacquers, nail varnish etc	moderately toxic if inhaled, swallowed or absorbed through the skin. The vapour can be irritating. Repeated exposure can cause damage to the eyes

Ethyl alcohol *see* ethanol

Ethyl glycol acetate	cellulose paints	moderately toxic if swallowed, inhaled or

		absorbed through the skin, also a skin irritant
Ethylene glycol	anti-freeze	toxic if swallowed and a possible skin irritant
Isopropanol	windscreen de-icers	moderately toxic if swallowed or inhaled. Dangerous to the eyes

Isopropyl alcohol *see* isopropanol

Methanol	de-icing mixtures and paint strippers	highly toxic if swallowed, also toxic by inhalation or absorption through the skin. Damages optic nerves by acute or chronic doses and can cause blindness in extreme cases

Methyl alcohol *see* methanol

Methyl ethyl ketone	general solvent	moderately toxic if inhaled, also a mild skin irritant

Methyl chloroform *see* 1, 1, 1 trichloroethane

Methylene chloride *see* dichloromethane

Naphtha	paint brush restorers etc	moderately toxic and also narcotic if inhaled
Styrene	2-part fillers etc	highly toxic if swallowed, moderately toxic if inhaled. It is a skin irritant, and the vapour can also irritate the eyes. Chronic exposure can damage the central nervous system. Mutagen and suspected carcinogen. However, risks less than they appear at first sight because not very

volatile, and hence contamination through breathing not likely

Tetrahydro-furan	general solvent	moderately toxic if swallowed or inhaled, also a narcotic. Irritating to eyes and breathing. Chronic exposure may damage liver and kidney

THF *see* tetrahydrofuran

Toluene	glues and damp treatments	moderately toxic if inhaled. Mutagenic and an eye and skin irritant
1, 1, 1 tri-chlorethane	general solvent	moderate to low toxicity. It is a possible carcinogen, but the evidence remains uncertain. Releases toxic phosgene gas when burnt. An important oxone depleter which is not yet covered by any international protocols for control
Trichloro-ethylene	DIY (Do It Yourself) products	highly toxic if inhaled with both immediate and long-term health effects. It is a mutagen, teratogen and carcinogen, and used to be an important component of smog in some US cities. Has been restricted in the United States

Trike *see* trichloroethylene

Turpentine	general solvent, used in paint removers etc	highly toxic if inhaled and chronic exposure can cause damage to the central

		nervous system. Possible carcinogen. Irritating to skin and mucous membranes
Turpentine substitute	paint and brush cleaner etc	variable composition so difficult to predict. Likely to be toxic if swallowed or inhaled. Can be carcinogenic, depending on composition
White spirit	in paints and paint brush cleaners	moderately toxic if swallowed or inhaled. Prolonged exposure can cause brain damage; also a possible carcinogen
Xylene	general solvent	moderately toxic if inhaled or absorbed through the skin. Chronic exposure can cause effects to the central nervous system

Assessment

The table makes interesting reading. Flick through and you'll find that some of the oldest solvents (white spirit, turps, trike) are actually amongst the most dangerous. It is difficult to tell whether this is because they have been studied more carefully, or because the older materials were really more hazardous.

Solvents certainly aren't very healthy materials to use a lot of. It is worth avoiding skin contact wherever possible and also avoiding breathing them continuously in a confined space. *Many are highly inflammable, so never use them near a naked flame.*

Further Information

There are a large number of directories of chemical and chemical hazards available. Try, for example, *The Risk Assessment of Environmental and Human Health Hazards: A Textbook of Case Studies* edited by Dennis J. Paustenbach (Wiley).

TAMPONS

The Issue
Tampons have been associated with risks from *toxic shock syndrome, dioxins* (see page 211) and ulceration.

The Facts
The membranes of the vaginal passage are thought to be very absorbent and sensitive to toxic substances. A growing number of specialists believe that contaminants in tampons can be absorbed into the body and cause both immediate and long-term health problems. Other people disagree, and say that absorption through the vagina will be minimal, especially during menstruation.

There are two main health issues: *Toxic shock syndrome* (TSS) is a serious disease which is commonest in women using tampons. Roughly 5 per cent of the 2,000 odd cases reported in the USA have resulted in death, although health experts believe that there is substantial under-reporting. Ninety cases had been reported in Britain by the end of 1984.

We still know relatively little about the disease. It has been suggested that the bacteria *Staphylococcus aureus* is the actual transmitter, but this hasn't been proved. It seems fairly indisputable that tampons increase the risk of TSS developing and there is some fairly inconclusive evidence that those made with artificial materials increase the risks even further. Rayon, which could well be contaminated with *dioxins* (see page 211) is a common material. On the other hand, the cotton used in tampons is likely to have had heavy pesticide applications and may well contain residues.

Countries vary a great deal as to how seriously they take TSS. The United States experienced a large drop in sales of tampons after a lot of media coverage in the 1980s. US companies now voluntarily list major components of the tampons on the box. Japan has had standards in place since 1969. In Britain, tampons fall through a legal loophole, being classified as neither medicines nor cosmetics, and aren't controlled. Although companies have a voluntary code of practice, standards are not as stringent as in some other countries.

Dioxins in tampons are an altogether more contentious isue. A report by the London-based Women's Environmental Network in 1989 caused an uproar from the industry. There is still a debate about whether dioxins can be transferred across the vaginal tisues or not, and about the risks involved in the concentrations found in tampons. Health risks at various levels are discussed in the section on *dioxins* (see page 211).

Ulceration is caused by drying of the vaginal tissues and is worst when tampons are used between periods to stop spotting, and for very light periods. Cracking is uncomfortable and also opens the area for further infection by contaminants of the tampon, and general viral and bacterial pathogens. There have been reports from the United States that regular use of tampons actually lengthens the period by increasing the bleeding. The Food and Drug Administration in the United States are now suggesting that more detailed absorbency ratings are listed on packages to allow women to choose the minimum absorbency which will be effective. In most countries there are just very broad categories given on the packets.

Assessment
It does seem likely that using tampons increases risks of Toxic Shock Syndrome, and may increase the chance of being contaminated by *dioxins*. On balance it looks as if tampons made from natural fibres are slightly less risky, but there isn't any firm proof of this as yet. However, risks of TSS are generally fairly small compared with the convenience of a tampon, and sanitary towels are not free from dioxin contamination either, although the potential for entry into the body is far less.

Minimizing Risks

- There are a few unbleached tampons available (which will have far less dioxin) and these are worth hunting out if you are worried;
- Choosing tampons of the correct (i.e. minimum possible) absorbency will reduce risk of ulceration;
- Cotton material may be less contaminated by dioxins (but may itself harbour pesticide residues);
- If you are really worried, panty liners are believed to be a less risky alternative.

Further Information
The Women's Environmental Network in London have carried out a campaign on the whole issue of dioxins in paper products

and have published an excellent book *The Sanitary Protection Scandal* by Alison Costello, Bernadette Vallely and José Young.

WASTE TREATMENT PLANTS

The Issue
Waste treatment plants dispose of *toxic waste*, usually by burning (incineration). Are these plants a hazard to health? Do people living nearby have cause to wory?

The Facts
Unfortunately 'the facts' are not really known in this case. Environmental groups are currently split about whether waste treatment plants are the best solution to toxic waste or not; Greenpeace is opposed to incineration for example, while other organizations say that it is a better option than either dumping or indefinite storage of hazardous wastes. And despite the huge controversy about waste plants in general, there are no studies of overall effects on either wildlife or humans.

The main issues relating to human health are:

- Risks of spillage of hazardous wastes on the way to the waste treatment plant, resulting in either contamination of drinking water sources or direct contamination of soil or air;
- Release of large amounts of toxic materials in the case of an accident;
- Similar releases as the result of deliberate and illegal 'overloading' of the system to burn more waste. (Many waste treatment plants are privately run, and profits are related directly to the amount of waste – particularly toxic waste – incinerated);
- Regular releases of toxic materials into the atmosphere through routine burning of waste.

A great deal of unpleasant material can come out of a waste treatment plant, including *dioxins* (see page 211), *radioactive material* (see page 231), *pesticides* (see page 173) (which sometimes become more toxic when burnt), solvents (see page 236) and heavy metals. Issues relate to how much and how dangerous. Some waste materials remain unaltered (e.g. arsenic, chromium and lead) while others are not completely destroyed by the incineration process. There are also 'new' pollutants formed by burning; these are called products of incomplete combustion (PICs).

The British Department of the Environment agrees that over 400 other products of incomplete combustion will be emitted from waste treatment plants.

In Britain there has been considerable debate about waste treatment plants in Bonnybridge in Scotland; Pontypool in Wales, and an incinerator which has now been closed near Nottingham. Despite considerable research, no firm evidence of health effects was found.

One of the key factors is how carefully the plant is run in practice. While there are many honest and careful waste treatment companies, there is a depressing list of others which have flaunted regulations and overloaded systems, taken inadequate care of controls and suffered continual accidents and spillages.

Assessment
We still don't have any real answers. If you live near an incinerator, there is enough information about harmful emissions to justify some concern. The greatest dangers lie in the plant being operated badly, so it is worth trying to keep a watching brief on safety standards if at all possible.

Minimizing Risk
One opportunity for countries belonging to the European Community is that there are stiff new regulations about incineration, including that of waste plants run by local authorities. Many small plants do not currently meet these targets, and there will doubtless be attempts in some areas to continue these practices for as long as possible. If you are worried, check that your local plant actually meets the new laws.

Further Information
Information on waste treatment plants and their problems can be obtained from Greenpeace offices around the world. Inform (381 Park Avenue South, New York, NY 10016; (212) 689-4040) has published a large number of detailed reports about the toxic waste issue in the United States.

WOOD PRESERVATIVES

The Issue
Many people use pesticides to treat timber in their homes. A

number of the pesticides cleared for timber treatment, especially in Britain, are poisonous, and some are among the most toxic known. There have been several cases of fatalities, or serious injuries, which seem likely to have been caused by timber treatment.

The Facts

Over the last few years, it has become much more common to treat timber against infestations of woodworm, dry rot, wet rot and other insect and fungal attack. Firms offer very tempting terms; promises of a guarantee of thirty years are by no means uncommon in Britain, although are probably very difficult to claim on in practice.

There are a number of serious problems with timber treatment by pesticides. The pesticides used are, by their nature, extremely persistent. Several of those commonly used in Britain, for example, have been banned in many other countries. They are dangerous to people and have had catastrophic effects on populations of bats in Britain, which have relied on nesting in eaves of roofs for hundreds of years. There are also serious doubts about the effectiveness of the treatments, and their necessity in buildings which are properly maintained.

Some of the most hazardous pesticides known are routinely used for timber treatment. They can be absorbed into the body through breathing contaminated air, especially in the initial period after treatment, although sometimes for a very long time afterwards as well. They can also contaminate through ingestion after touching treated timber, or straight through the skin. People working in roof spaces after timber treatment can get high doses through touching the timber and breathing air in a confined space, although fumes can also reach into the main body of the building as well.

The table below lists some of the commonest pesticides used in timber treatment, along with their effects on human health. Note that many of the pesticides listed under *garden pesticides* (see page 179) also turn up in timber treatment; if you don't find a pesticide listed below, try the other section as well.

Pesticide	Uses	Effects
Alkyl-ammonium	DIY and professional products	Little known about toxicity

Arsenic	In some pre-treatments as an insecticide and fungicide, usually with copper and chrome	A deadly poison – poisoning has occurred through, handling wet timber. Splinters tend to fester painfully. A chronic poison with a wide range of effects, including carcinogenicity and teratogenicity. Several different arsenic compounds are used in timber treatments
Boron compounds	Fungicides and insecticides used as water soluble rods and pellets inserted into drilled spaces, and as a pre-treatment preservative of green timber	As a pretreatment, probably one of the least hazardous as insecticides available, although it is a poison and deaths have occurred by accidental swallowing. Little known about chronic toxicity
Creosote	Has been commonly used in timber treatment as a fungicide and insecticide, although most often used for fencing and garden furniture. Despite still being a widely used DIY product in Britain, it has been banned for all but professional use in the United States	Highly poisonous, causing skin and eye irritation with the possibility of permanent damage to the cornea of the eye. Acute bronchitis can be brought on in people caught in the spray mist, along with nausea and headache. Carcinogenic
Dieldrin	Insecticide used in some wood	Extremely toxic: any wood preservative containing, this

preservatives. Banned in the United States since 1975 and should be finally banned for all uses in Britain by 1992

is probably already illegal, and certainly shouldn't be used

Gamma HCH *see* lindane

Lindane
A highly poisonous insecticide used in many timber treatments in Britain and the United States, although banned in some countries

Acts as an irritant, by inhalation and through the skin. Extremely persistent and has been suspected of causing nervous disorders, including the onset of epilepsy. Can also cause damage to the blood system and is strongly suspected of being a carcinogen. Implicated in illness and death in several people after their homes had been treated against various forms of fungal disease. Highly poisonous to wildlife as well and can remain lethal to bats for more than twenty years after timber treatment

PCP *see* pentachlorphenol

Pentachlor-
phenol
A fungicide used in pre-treatment or remedial treatment, and available in many DIY formulations, sold over the counter in Britain. In the United

Highly toxic poison, which has been blamed for over a thousand deaths worldwide. Extremely persistent, and a contact poison which acts through the skin. Can affect nerves, kidney, liver and heart, and can cause chloracne. Can also contain

States it has been banned for home consumer use since July 1984, and there are strict controls on professional use

contaminants, such as dioxins, which are teratogenic

Permethrin

An insecticide which is generally considered to be far safer to bats and other wildlife and is recommended as a safe alternative by the British Nature Conservancy Council (a government body) in their advice leaflets about protecting bats

Has been the subject of acrimonious dispute in the United States, and some senior researchers at the Environmental Protection Agency believe that it is a powerful carcinogen in humans. Dr Adrian Gross, a researcher for the EPA, resigned in protest at what he believed was a cover-up of permethrin's carcinogenicity, but no controls have been set on its use as yet

TBTO, Tributyl tin oxide

Important fungicide in many British timber treatments

A highly toxic fungicide which is also an irritant and can be absorbed through the skin. Suspected teratogen, since the US EPA found it to cause birth defects in animals in 1986. Has been banned for use in anti-fouling paint in Britain because of its impact on commercial mussel (shellfish) beds. Used for pre-treatment and remedial treatment of timber and widely available from DIY and timber merchant shops

In January 1989, the London Hazards Centre published a book,

Toxic Treatments, which examined a number of case studies of people affected by poisoning as a result of domestic timber treatment. One of the cases was of Eric Riley who died in January 1988, a year after professional timber specialists treated his house with pentachlorphenol and lindane. Eric and his wife Ann felt ill throughout the year following the treatment and Eric suffered two epileptic fits. Unfortunately, the second one took place while he was in the bath, and he drowned. The Coroner recorded an open verdict. In 1989, over a hundred British MPs signed an Early Day Motion demanding the banning of lindane and pentachlorphenol (PCP), although the initiative was rejected by the government.

Assessment
Timber treatment by powerful insecticides is certainly hazardous to some people, although whether this is because a minority are particularly sensitive, or they are simply unlucky enough to be badly contaminated is still not known.

In most cases remedial timber treatment with powerful pesticides is unnecessary. Buildings maintained in good condition, with a damp-proof course to prevent rising damp, and at an adequate temperature, will not usually develop problems of insecticidal or fungal attack. Poor building structure is often to blame for susceptibility to attack and there is also a great deal of persuasive advertising by firms. Treatment is by no means always necessary.

Minimizing Risk
Only use pesticides on timber if you think it is absolutely necessary. And if you do use them, avoid the most hazardous, i.e. those which contain arsenic, lindane or pentachlorphenol. This won't give you complete protection from risk of course; all pesticides are dangerous.

Pesticides are unlikely to appear in the name of the product, but should be listed on the label. If you employ professional timber treatment 'experts' make sure that they are not using these chemicals; some of the most toxic are allowed for professional use even in countries which don't license them for use at home. But you still have to live with the fumes afterwards.

Further Information
The best book currently available on timber treatment hazards, which includes a much more detailed summary of hazards from

different chemicals than I have space for here, is *Toxic Treatments* from the London Hazards Centre, Headland House, 308 Grays Inn Road, London WC1X 8DS.

X-RAYS

The Issue
X-rays are an accepted part of medical diagnosis. They expose the body to radiation. Are they worth the risk?

The Facts
There are four main uses of radiation in medicine:

- Diagnostic radiology: the use of X-rays to see what is happening in various parts of the body;
- Dental radiography: the use of X-ray pictures of the teeth and jaw, treated separately because X-rays are normally carried out in the dentist's own surgery;
- Nuclear medicine: the use of radionuclides in diagnosis;
- Radiotherapy: use of ionizing radiation in the treatment of cancer.

Radionuclides and radiotherapy are only used in serious medical cases, when exposure to radiation is judged to be the lesser of two evils. (Their effectiveness, or otherwise, is not our concern here.) However, full body and dental X-rays are used as a standard and constant diagnostic technique by doctors in treating even quite minor injuries and ailments. Other, more highly radioactive, procedures are also used, such as barium meals and barium enemas.

These figures are obviously very approximate. But they are worrying enough for B. W. Wall, of the NRPB, to write:

> The assumption of trivial level of risk ... is seen to be invalid for most types of examination. In particular, examinations involving the irradiation of the lower abdomen of pregnant women are seen to require extremely high levels of justification and optimization.

This is probably not yet the end of the story. Recent changes in estimates of the health effects of radiation generally, based on new evidence from the health effects of the Chernobyl disaster, suggest that figures for health effects from X-rays will continue to be reduced. Keep an eye open for further developments in this field.

Assessment

Risks of developing cancer from X-rays are probably very small in most cases. But they are not zero. It is always worth questioning whether an X-ray is strictly necessary and, if you think that it is not (for example at the dentist when X-rays are increasingly being used to, double check on analysis), then you are within your rights to refuse one. Certainly, pregnant women should avoid having X-rays unless these are strictly necessary.

What Governments Could Do

Governments could reduce risks in three main ways:

- By providing information about risk levels to doctors, dentists and to the public, so that patients are able to make informed decisions about whether they want X-rays or not. This is especially important in situations where the person carrying out the treatment charges for the X-ray, and therefore has a direct interest in using the equipment;
- By ensuring that X-ray equipment is maintained adequately, and used properly, through tightening up on legislation concerning radiography;
- In countries with some kind of national health service, better investment in radiography can improve safety margins and reduce risk to patients.

Further Information

The book *Radiation Risks: An Evaluation* by David Sumner (Tarragon Press, 1987) gives a good basic introduction to the issues and the debates.

A collection of technical papers from the Institute of Physical Sciences in Medicine are also useful: *Are X-rays Safe Enough? Patient Doses and Risks in Diagnostic Radiology* editied by K. Faulkner and B. F Wall, Institute of Physical Sciences in Medicine Report number 55, 1988 (published in Britain).

The National Radiological Protection Board have also produced guidelines for pregnant women; *Exposure of ionising radiation to pregnant women: Advice on the diagnostic exposure of women who are, or may be, pregnant,* NRPB ASP8, (HMSO, London).

AFTERWORD

This book was written to try and make sense of some of the most noteworthy environmental health issues brought to public attention at the beginning of the 1990s. It makes no claim to being a complete guide to 'enviromental hazards'; but answers the main questions that people are likely to be asking at the moment. It is also, I am sure, influenced by my own particular interests and obsessions. Future editions will doubtless contain new categories, as other issues come into prominence.

To repeat what was written in the introduction, *don't use this book as a blueprint for paranoia*. There *are* a lot of worrying environmental pollutants about. You *can* often take simple steps yo avoid them. But it is worth remembering that on average human lifespan is still getting longer, and a brief encounter with a possible carcinogen in dilute form does not mean you have to suffer nightmares as a result!

The editors of this book initially encouraged me to try and put 'hazard ratings' against the various entries. I resisted, simply because in most cases we have very little idea of the level of risk. The pesticide industry complains, with some justification, that it gets a worse press than many other hazardous industries simply because more research is carried out on agrochemicals, and they are more in the public eye. Perception of the risk from radon and X-rays altered substantially even as this book was being drafted, showing how poor our understanding remains in many areas. There are still many areas where we have only the most hazy understanding.

We are a long way from being able to compare the risk between say, accidental exposure to solvent fumes and walking through a field of bracken in the autumn. Attempts to rank risks should therefore be taken with a pinch of salt. The constantly changing assessment of the scale of risks from low-level radiation are a case in point. At each stage of its development the nuclear industry has argued powerfully, and convincingly, that their risk analysis was accurate, questionably and, if anything, over-cautious. Inevitably, further research has been carried out and the figures have changed. Absolute certainty is claimed again, until the next change.

Hazard watchers are also in danger of overreacting to occasional scare stories in the media or from well meaning individuals and organizations. These need to be treated with caution as well. A number of newspaper articles in recent years have substantially over-emphasized a particular risk.

Taking Steps to Protect Ourselves

There are now some common threads appearing in research, and from people trying to give advice about environmental hazards. First, many pollutants which are dangerous, or potentially dangerous, work in concert with other pollutants, or are more likely to be dangerous to someone in poor health, with a reduced immune system, and so on. Smoking, drinking alcohol, a poor diet, lack of exercise and other traditionally unhealthy pastimes are therefore likely to increase the risk of an individual suffering after exposure to other toxic materials. In other words, if you live a healthy life with a good diet you are *also* helping protect yourself.

Second, *repeated* exposure to most of the things described above is much, much more dangerous than the occasional dose, unless the latter is very large or prolonged. And children (especially infants) are likely to be much more susceptible to many of these pollutants than healthy adults.

Third, in many cases there are things that we can do, quite easily, to reduce levels of risk. This doesn't mean that it is *only* the individual's responsibility. In most of the issues described above, there are clear steps which governments and industry can take to reduce risks; I have sometimes summarized these. But in the short term, informed decisions about which products to use, or care in avoiding the more hazardous situations, can make a real difference.

Fourth, there already is quite a lot of information available on at least some of the issues discussed in this book. *Always read the label* when you are using something which is known to be, or thought to be, dangerous. For example, pesticide packets will contain information about how long to leave crops before eating them; solvents will often say if they are acutely poisonous. There are also some international hazard symbols (illustrated below) which are worth learning for quick reference. However, remember that not everything is adequately labelled. As we have seen, there are now laws requiring identification of the contents of many household or cosmetic products in most countries, so just because something *doesn't* have anything on the label is no guarantee that it is completely non-toxic.

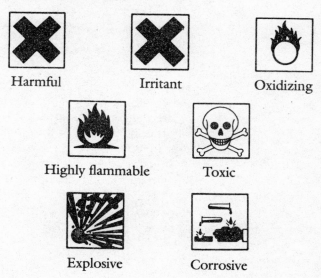

Harmful   Irritant   Oxidizing

Highly flammable   Toxic

Explosive   Corrosive

Finally, and perhaps most usefully, there are an increasing number of sources within organizations providing information and advice for people worried about specific environmental hazards. These range from trades union safety reps, through advice given by manufacturers, to specialized organizations dealing with a particular hazard. In the case of small non-governmental organizations, however, you should expect to pay something for information which is provided, since their funds and staff are likely to be very limited.

Use this book wisely, but not obsessively! And stay healthy.

Nigel Dudley
Machynlleth, September 1990

SOURCES OF HELP
AND INFORMATION

This book has raised a lot of questions. In covering so many different subjects, it isn't always possible to go into things in the depth that they deserve. Contacts for more information have been given for several countries at the bottom of each section, where relevant. However, there are also a number of organizations whose interests span more than one of the subjects listed here, and details of these are given below.

A note of caution. Many of the groups listed are non-governmental organizations, a lot are small, underfunded and understaffed. Please don't expect miracles. *Any* letter should include a stamped addressed envelope if you want a reply, and ideally a donation if you are asking for any detailed information. Many groups will simply reply with standard literature, if at all. Bear with them. Anyone who has had a daily postbag of enquiries will know the enormous demands in answering individual queries.

Likewise, when phoning, bear with people if they ask you to ring back later because they are snowed under. *Never* expect anyone to ring you back on their phone bill. And again, a donation is in order if anyone spends a long time giving you information. Environmental groups, health and safety groups and others daily give away information which a commercial consultant would charge tens or hundreds of pounds for.

GROUPS IN AUSTRALIA

Albury-Wodonga Environment Centre: Shop 3, 608 Dean Street, Albury NSW 2640

Australian Conservation Foundation: GPO Box 1875, Canberra ACT 2601

Barrier Environment Group: West Post Office, Broken Hill NSW 2880

Cairns and Far North Environment Centre: 1st Floor, Andrejic Arcade, 55 Lake Street, Cairns QLD 4870

Canberra and South East Region Environment Centre: Childers Street Buildings, Kingsley Street, Acton ACT 2601

Capricorn Conservation Council and Environment Centre: 135 William Street, Rockhampton QLD 4700

Central Australian Conservation Council: PO Box 2796, Alice Springs, NT 5750

Conservation Council of Victoria: 247-251 Flinders Lane, Melbourne VIC 3000

Conservation Centre and Conservation Council of SA Inc: 120 Wakefield Street, Adelaide SA 5000

Conservation Council of the South East Region and Canberra: GPO Box 1875, Canberra ACT 2601

Conservation Council of WA Inc.: 1st Floor, 794 Hay Street, Perth WA 6000

Friends of the Earth: 366 Smith Street, Collingwood, VIC 3066

Illawarra Environment Centre: Suite 11, 157 Crown Street, Wollongong NSW 2500

Launceston Environment Centre Inc.: 34 Patterson Street, Launceston TAS 7250

Mid North Coast Environment Centre: c/o 'Coolenberg', 60 Lake Road, Port Macquarie NSW 2444

Nature Conservation Council of NSW: NSW Environment Centre, 176 Cumberland Street, Sydney NSW 2000

Newcastle Ecology Centre: Room 6, Trades Hall, Union Street, Newcastle NSW 2300

North Coast Environment Council: Pavans Road, Grassy Head, via Stuarts Point NSW 2441

North Queensland Conservation Council Inc: 477 Flinders Road, Townsville QLD 4810

South Coast Environment Centre: Shop 5, Sams Market Arcade, Bateman's Bay NSW 2500

Tasmanian Conservation Trust: 102 Bathurst Street, Hobart, TAS 7000

The Environment Centre of NSW: 176 Cumberland Street, Sydney NSW 2000

Total Environment Centre: 18 Argyle Street, Sydney NSW 2000

World Wide Fund for Nature (Australia): Level 17, St Martins Tower, 31 Market Street, GPO Box 528, Sydney, NSW 2001

GROUPS IN EUROPE

Bureau Européen des Unions de Consommateurs: 29 Rue Royale Bte 3, 1000 Brussels, Belgium; 2-218-3093

European Environment Bureau: 29 rue Vautier, 1040 Brussels, Belgium; 322-647-0199

European Consumers Consultative Committee: 200 Rue de la Loi, B-1040, Brussels, Belgium

Greenpeace International: Keizerstrasser 176, 1012 DW Amsterdam, Netherlands

Institute for European Environmental Policy: Aloys-Schulte Strasse 6, D-5300 Bonn 1, Federal Republic of Germany 49-228/21 38 10

International Organization of Consumers Unions: 9 Emmastraat, 2595 EG, The Hague, Netherlands; 70-476-331

Pesticide Action Network: PAN-Europe, 22 rue des Bollandistes, 1040 Brussels, Belgium; 2-734-2332

Pesticide Action Network: Damrak 83-1, 1012 LN Amsterdam, Netherlands

World Wide Fund for Nature: Avenue du Mont-Blanc, CH 1196 Gland, Switzerland; 022-649-111

GROUPS IN MALAYSIA

International Organization of Consumer Unions (IOCU): PO Box 1045, 10830 Penang, Malaysia

Sahabat Alam Malaysia (Friends of the Earth Malaysia): 37 Lorong Birch, 10250 Penang, Malaysia

World Rainforest Movement International Secretariat: 87 Catonment Road, 10250 Penang, Malaysia

GROUPS IN NEW ZEALAND

Friends of the Earth: PO Box 39-065, Auckland West, New Zealand

Greenpeace: Nagel House, 5th Floor, Couthouse Lane, Auckland, New Zealand

World Wide Fund for Nature: PO Box 6237, Wellington, New Zealand

GROUPS IN THE UK

Advisory Committee on Pollution of the Sea (ACOPS):

3 Endsleigh Street, London WC1H 0DD; 071-388-2117

Ash (Action on Smoking and Health: 5-11 Mortimer Street, London W1N 7RH; 071-637 9843

British Effluent and Water Association: 51 Castle Street, High Wycombe, Bucks HP13 6RN; 0494 444544

British Medical Association: BMA House, Tavistock Square, London WC1H 9JP

British Waterways Board: Llanthony Warehouse, Gloucester Docks, Gloucester, GL1 2EH; 0452 25524

Compassion in World Farming: Lyndham House, Greatham, Petersfield, Hampshire GU32 3EW; 0730 64208

Consumers' Association: 2 Marylebone Road, London NW1 4DX

Consumers in the European Community: 24 Tufton Street, London SW1P 3RB; 071-222-2662

Department of the Environment: 2 Marsham Street, London SW1

Earth Resources Research: 258 Pentonville Road, London N1 9JY; 071-278-3833

Food Additive Campaign Team (FACT): c/o Erik Millstone, Mantell Building, University of Sussex, Brighton

Food Commission: 88 Old Street, London EC1 9AR. 071-253-9513

Friends of the Earth International: 26 Underwood Street, London N1; 071-490-1555. Also *Friends of the Earth Scotland*: 15 Windsor Street, Edinburgh 7, 031-557-3432

Green Alliance: 60 Chandos Place, London WC2N 4HG; 071-836-0341

Greenpeace: 30-31 Islington Green, London N1; 071-359-7396

Hazards Bulletin: Sheffield, PO Box 148, S1 1FB

Institute of Environmental Health Officers: Chadwick House, Rushworth Street, London SE1 0QT

London Hazards Centre: 308 Grays Inn Road, London WC1

Marine Conservation Society: 4 Gloucester Road, Ross on Wye, Herefordshire, HR9 5BU; 0989-66017

National Anti-Fluoridation Campaign: 36 Station Road, Thames Ditton, Surrey KT7 0NS; 081-398-2117

National Consumers Council: 20 Grosvenor Gardens, London SW1W 0DH; 071-636-4066

National Poisons Information Service: New Cross Hospital, Poisons Unit, Avoney Road, London SE14 5ER; 071-407-7600 (emergencies only: 071-635-9191)

National Pure Water Association: Bank Farm, Aston Piggot, Westbury, Shrewsbury, SY5 9HM; 074-383-445

National Society for Clean Air: 136 North Street, Brighton, Sussex BN1 1RG; 0273 26313

Parents for Safe Food: Britannia House, 1-11 Glenthorpe Road, Hammersmith, London W6 0LF; 081-748-9898

Pesticide Exposure Group of Sufferers (PEGS): 10 Parker Street, Cambridge

Pesticide Trust: 23 Beehive Place, London SW9

Ramblers Association: 1-5 Wandsworth Road, London SW8 2LJ; 071-528-6878

Scottish Consumers Council: 314 St Vincent Street, Glasgow G3 8XW; 041-226-5261

Scottish River Purification Boards Association: City Chambers, Glasgow, G2 1DU; 041-221-9600

Soil Association: 86 Colston Street, Bristol, BS1 5BB; 0272-290661

Transport 2000: Walkden House, 10 Melton Street, London NW1 2EJ; 071-388-8386

UK Genetics Forum: 3-4 St Andrews Hill, London EC4V 5BY

VDU Workers' Rights Campaign: c/o City Centre, 32-35 Featherstone Street, London EC1; 071-608-1338

Water Authorities Association: 1 Queen Annes Gate, London SW1 9BT; 071-222-8111

Water Research Centre: 660 Ajax Avenue, Slough, Berks; 0753-37277

Welsh Consumers Council: (*Cygnor Defnyddwyr Cymru*), Castle Buildings, Womanby Street, Cardiff; 0222-396056

Women and Work Hazards Group: c/o A Woman's Place, Victoria Embankment, London WC2N 6PA

Women's Environmental Network: 287 City Road, London EC1V 1LA; 071-490-2511

World Wide Fund for Nature: Panda House, Weyside Park, Godalming, Surrey GU7 1XR; 0483-426444

GROUPS IN THE USA

Acid Rain Foundation: 1410 Varsity Drive, Raleigh, NC 27606; (919) 828-9443

CARE for the Earth: 41 Sutter Street, Room 300, San Francisco, CA 94104; (415) 781-1585

Centre for Science in the Public Interest: 1501 16th Street NW, Washington DC 20036; (202) 332-9110

Citizens for a Better Environment: 942 Market Street, #505, San Francisco CA 94102; (415) 788-0690

Citizens Clearinghouse for Hazardous Wastes: PO Box 926, Arlington, VA 22216; (703) 276-7070

Clean Water Action Project: 317 Pennsylvania Avenue SE, Washington DC 20003, (202) 547-1196

Concern Inc: 1794 Columbia Road NW, Washington DC 20009; (202) 328-8160

Conservation Foundation: 1250 24th Street NW, Washington DC 20037; (202) 797-4300

Conservation International: 1015 18th Street NW, Washington DC 20036; (202) 429-5660

Earth First: PO Box 5871, Tucson AZ 85703; (602) 622-1371

Earth Island Institute: 300 Broadway, Suite 28, San Francisco CA 94133; (415) 788-3666

Earthwatch: 1228 1/2 31st Street NW, Washington DC 20077; (202) 342-2564

Environmental Action Foundation: 1525 New Hampshire Ave NW, Washington DC 20036; (202) 745-4870

Environmental Defense Fund: 1616 P Street NW, Washington DC 20036; (202) 387-3500

Environmental Law Institute: 1616 P Street NW, Washington DC 20036

Environmental Policy Center: 218 'D' Street SE, Washington DC 20003; (202) 544-2600

Friends of the Earth: 218 'D' Street SE, Washington DC 20003; (202) 544-2600

Food First: 1885 Mission Street, San Francisco CA 94103

Global Tomorrow Coalition: 1325 G Street NW, Suite 1003, Washington DC 20005; (202) 628-4016

Greenpeace: 1611 Connecticut Avenue NW, Washington DC 20009; (202) 462-1177

National Recycling Coalition: 17 'M' Street NW, Suite 294, Washington DC 20036; (202) 659-6883

National Toxics Campaign: 37 Temple Place, 4th Floor, Boston MA 02111; (617) 482-1477

Natural Resources Defense Council: 1350 New York Avenue NW, Washington DC 20003; (202) 949-0049

The Nature Conservancy: 1815 N Lynn Street, Arlington, VA 22209; (703) 841-8737

Pesticide Action Network: 965 Mission Street, San Francisco, CA 94103; (415) 771-7327

Rachel Carson Trust: 8940 Jones Mill Road, Chevy Chase, MA 20815

Resources for the Future: 1616 'P' Street NW, Washington DC 20036; (202) 328-5000

The Sierra Club: National Office, 730 Polk Street, San Francisco, CA 94109; (415) 776-2211

Water Information Network: Box 909, Ashland, OR 97520; (800) 533-6714

World Resources Institute: 1750 New York Avenue NW, Washington DC 20006

World Wide Fund for Nature (USA): 1250 24th Street NW, Washington DC 20037; (202) 293-4800

Worldwatch Institute: 1776 Massachusetts Avenue, Washington DC 20036; (202) 452-1999

RECOMMENDED READING

This is very select! I have listed key books at the end of each section. Here is a brief listing of some of the more general textbooks of environmental hazards which readers may want to refer to in following up specific chemicals, or in finding out information on topics not covered by this book.

Dudley, Nigel (1986), *Garden Pesticides*, The Soil Association, Bristol

Paustenbach, Dennis J. (1989), *The Risk Assessment of Environmental and Human Health Hazards: A Textbook of Case Studies*, Wiley

Sax, N. Irving (editor), *Dangerous Properties of Industrial Materials*, van Nostrand Reinhold (frequently updated)

Watterson, Andrew (1988), *Pesticide Users Health and Safety Handbook*, Gower

There are a large number of directories and encyclopedias of chemical hazards; any decent sized reference library should have a few. These are always frustrating to use, because none can ever contain everything, but in looking through everything available you should be able to build up an informed overview.

GLOSSARY

Most of the terms used in this book are defined in the text or the introduction. The following glossary explains some of the acronyms which are used in the text, or which you may come across if you read around the various subjects.

ADI: Acceptable Daily Intake. The amount of a toxic material defined as the maximum which can be ingested every day without suffering ill effects
AMA: American Medical Association
BMA: British Medical Association
BNFL: British Nuclear Fuels Ltd
BRE: Building Research Establishment (UK organization)
CMHR: Combustion Modified High Resilient Foam. Foam which has been proofed against fire
EC: European Community
ELF: Extremely Low Frequency radiation
EPA: United States Environment Protection Agency
FAO: Food and Agriculture Organization of the United Nations
FDA: Food and Drug Administration of the United States
HDRA: Henry Doubleday Research Association (UK non-governmental organization)
HSE: Health and Safety Executive (part of UK government)
IAEA: International Atomic Energy Association (UN body)
IFOAM: International Federation of Organic Agriculture Movements (international non-governmental organization)
ILO: International Labour Organization of the United Nations

IOCU: International Union of Consumer Organizations (international non-governmental organization)

LD50:   amount of chemical which kills 50 per cent of laboratory animals tested

MMF: man-made fibres

MRL: Maximum Residue Limit. Maximum amount of pesticide permitted to remain as a residue on food

NAS: National Academy of Science (US body)

NFU: National Farmers' Union (UK trade union)

NIOSH: National Institute for Occupational Safety and Health (US body)

NRPB: National Radiological Protection Board (UK body)

NSCA: National Society for Clean Air (UK non-governmental organization)

NCC: Nature Conservancy Council (UK government body)

PAH: Polycyclic Aromatic Hydrocarbon

PAN: Pesticide Action Network (international non-governmental organization)

PEGS: Pesticide Users Exposure Group of Sufferers (UK non-governmental organization)

RCEP: Royal Commission on Environmental Pollution (UK based commission)

SBS: Sick Building Syndrome

TLV: Threshold Limit Value. Maximum aerial concentration of a pollutant allowed

TSS: Toxic Shock Syndrome

WHO: World Health Organization of the United Nations

WRI: World Resources Institute

INDEX